Wine Tourism and Sustainability

"This well-written textbook, edited by an experienced group of wine tourism researchers and teachers, fills a major gap in the current wine tourism textbook market. It does a commendable job of introducing core wine tourism concepts including economic impacts, sustainability, social issues, and competitiveness. However, its real advantage is the care the authors take to introduce cutting edge concepts that are relevant to identifying best practices for students of wine management, tourism and hospitality research, and industry leaders. This textbook is both interesting and very readable."

—Armand Gilinsky, *Emeritus Professor, Wine Business Institute, Sonoma State University*

"An essential journey through the world's vineyards, the book uncorks the essence of sustainability in the wine industry, revealing its economic, social, and environmental impact. A must-read pour for enthusiasts and experts alike."

—Gustav Visser, *Full Professor, Department of Geography & Environmental Studies, Stellenbosch University*

"Unveiling the rich tapestry of wine's sustainable journey, the book captures the wine tourism industry's dedication to economic, social, and environmental excellence. A captivating blend for those passionate about wine's deeper narrative."

—Nikos Georgantzis, *Full Professor, School of Wine & Spirits Business, Burgundy School of Business*

Javier Martínez-Falcó ·
Bartolomé Marco-Lajara ·
Eduardo Sánchez-García ·
Luis A. Millán-Tudela
Editors

Wine Tourism and Sustainability

The Economic, Social and Environmental Contribution of the Wine Industry

palgrave
macmillan

Editors
Javier Martínez-Falcó
Department of Management
University of Alicante
Alicante, Spain

Bartolomé Marco-Lajara
Department of Management
University of Alicante
Alicante, Spain

Eduardo Sánchez-García
Department of Management
University of Alicante
Alicante, Spain

Luis A. Millán-Tudela
Department of Management
University of Alicante
Alicante, Spain

ISBN 978-3-031-48936-5 ISBN 978-3-031-48937-2 (eBook)
https://doi.org/10.1007/978-3-031-48937-2

This Palgrave Macmillan imprint is published by the registered company Springer Nature Switzerland AG
The registered company address is: Gewerbestrasse 11, 6330 Cham, Switzerland

Paper in this product is recyclable.

Preface

In the annals of human history, few beverages have so deeply captured the imagination, culture, and traditions of societies as wine. Beyond the bottle, however, lies an equally compelling story: wine tourism. As travelers traverse the world in search of the perfect vintage, they embark on journeys that go beyond mere tasting, experiencing the rich tapestry of history, culture, and nature that each wine region has to offer. Against this backdrop comes *Wine Tourism and Sustainability—The Economic, Social, and Environmental Contribution of the Wine Industry*, published by the esteemed publishing house Palgrave Macmillan.

Wine tourism, a burgeoning facet of the global travel industry, is more than just a leisurely sojourn through picturesque vineyards. It is an exploration of history, a journey through the senses, and a celebration of the human spirit's ability to cultivate beauty from the earth. This book delves deep into the multifaceted world of wine tourism, unraveling its significance not just as a niche travel segment, but as a powerful force that contributes to the broader economic, social, and environmental landscape.

Economically, wine tourism is a beacon of prosperity, generating revenue for local businesses, creating employment opportunities, and fostering entrepreneurship. From quaint bed-and-breakfast establishments to large wine estates, the ripple effect of wine tourism affects several sectors, infuses vitality into local economies, and offers a sustainable model for growth. From a social standpoint, wine tourism builds bridges between cultures, as it brings together people from diverse backgrounds, fostering mutual respect, understanding, and appreciation. The shared experience of tasting, learning, and celebrating wine creates bonds that transcend borders, making the world a smaller and more connected place. It also preserves and promotes local traditions, ensuring that the rich tapestry of wine heritage is passed on to future generations. On the environmental side, wine tourism plays a key role in advocating sustainability. When travelers embark on journeys through vineyards and wine regions, they become ambassadors for the environment, often leaving with a heightened awareness of the delicate balance between nature and agriculture. Many wine tourism destinations prioritize environmentally friendly practices, from sustainable lodging options to tours that educate visitors about land conservation and biodiversity. This emphasis not only ensures that the beauty and health of these regions are preserved for future generations, but also instills in tourists a deeper appreciation for the environment.

Wine Tourism and Sustainability—The Economic, Social, and Environmental Contribution of the Wine Industry is more than just a book; it is an invitation. An invitation to explore the world of wine tourism, to understand its profound impact on our society, economy, and environment, and to celebrate the timeless allure of this magnificent beverage. As you turn the pages, may you be inspired by the stories, insights, and revelations that underscore the indomitable spirit of the wine tourism industry and its commitment to a sustainable future.

At a time when sustainability and responsible tourism have become of paramount importance, this book stands as a guiding light of insight and relevance. For academics, it fills a crucial gap in literature, offering a comprehensive exploration of wine tourism's multifaceted contributions to economic, social, and environmental sustainability. The rigorous analyses presented within these pages provide scholars with a robust foundation for further study and discourse. For industry professionals, whether they are winemakers, tour operators or policymakers, the book is a practical roadmap. It illuminates best practices, highlights innovative approaches to sustainability, and offers a roadmap for integrating these principles into their operations. Thus, *Wine Tourism and Sustainability—The Economic, Social, and Environmental Contribution of the Wine Industry* is not just an academic endeavor; it is a call to action that urges both academia and industry to recognize, embrace, and champion the transformative power of sustainable wine tourism.

Here's to a journey of discovery, reflection, and celebration. Cheers!

Alicante, Spain

Javier Martínez-Falcó
Bartolomé Marco-Lajara
Eduardo Sánchez-García
Luis A. Millán-Tudela

Acknowledgements

To José Falcó Martínez (1923–2023),

This book, although replete with academic wisdom about wine and wine tourism, could not have been completed without the inspiration that your life and legacy have given us. Like an aged wine, your warmth and kindness have left in us a taste that lingers, a memory that, with the passage of time, only becomes more precious.

Your sense of transcendence taught us that, beyond the tangible, there are values and moments that remain engraved in the soul, that transcend time and space. May this book not only be a tribute to the wine, but also to you, to your immense kindness and to the eternal imprint you have left in our hearts.

With deep respect and eternal love.

Contents

Contributors

Tim Baird Lincoln University, Christchurch, New Zealand

Dhruv Kishore Bole Faculty-Food and Beverage Service, State Institute of Hotel Management, Siddhpur, Gujarat, India

María del Carmen González-Velasco Department of Business Management and Economics, University of León, León, Spain

Pavel Castka UC Business School, University of Canterbury, Christchurch, New Zealand

Francesca Checchinato Agrifood Management and Innovation Lab, Department of Management, Ca' Foscari University of Venice, Venice, Italy

Cinzia Colapinto Agrifood Management and Innovation Lab, Department of Management, Ca' Foscari University of Venice, Venice, Italy

Vladi Finotto Agrifood Management and Innovation Lab, Department of Management, Ca' Foscari University of Venice, Venice, Italy

Rosana Fuentes-Fernández Department of Business Management and Economics, University of León, León, Spain

Marcos González-Fernández Department of Business Management and Economics, University of León, León, Spain

C. Michael Hall Department of Management, Marketing & Entrepreneurship, University of Canterbury, Christchurch, New Zealand

Galinoma Gahele Lubawa Institute of Rural Development Planning, Dodoma, Tanzania

Bartolomé Marco-Lajara University of Alicante, Alicante, Spain

Javier Martínez-Falcó University of Alicante, Alicante, Spain

Christine Mauracher Agrifood Management and Innovation Lab, Department of Management, Ca' Foscari University of Venice, Venice, Italy

Doris Miculan Bradley George Brown College, Toronto, ON, Canada

Luis A. Millán-Tudela University of Alicante, Alicante, Spain

Rosario Navalón-García Department of Regional Geographical Analysis, University of Alicante, Alicante, Spain

Evans S. Osabuohien Covenant University, Ota, Ogun State, Nigeria

Haywantee Ramkissoon UniSA International—University of South Australia, Adelaide, SA, Australia

Chiara Rinaldi Agrifood Management and Innovation Lab, Department of Management, Ca' Foscari University of Venice, Venice, Italy

Donna Lee Rosen Royal Roads University, Victoria, BC, Canada; SHTM, George Brown College, Toronto, ON, Canada

Eduardo Sánchez-García University of Alicante, Alicante, Spain

List of Figures

Developing Sustainable Tourism in Tanzania: Perspective from Wine Stakeholders

Spain's Wine Tourism: Evaluating the Economic Contributions of Wineries and Museums

Sustainable Wine Tourism: Best Practices

Wine Tourism, the Business of Wine, and the Impact of the Environöment and Sustainability

Wine Tourism in the Province of Alicante (Spain). Revitalization of Wine Heritage as a Strategy for Tourism Diversification and Differentiation in Mature Tourist Destinations

The Impact of Wine Tourism on the Sustainability Outcomes of Spanish Wineries: An Exploration via Structural Equation Modeling

List of Tables

Wine Tourism, Innovation, and Sustainable Winegrowing in Cool Climate Regions: A Longitudinal International Comparative Analysis

Economic Structure and Geographic Scope of Spanish Wine Routes

Developing Sustainable Tourism in Tanzania: Perspective from Wine Stakeholders

Spain's Wine Tourism: Evaluating the Economic Contributions of Wineries and Museums

Discovering *Enocotourism* in the Ribera del Duero Appellation: A New Concept Combining Wine Tourism and Ecotourism for a Unique Experience

Wine Tourism in the Province of Alicante (Spain). Revitalization of Wine Heritage as a Strategy for Tourism Diversification and Differentiation in Mature Tourist Destinations

The Impact of Wine Tourism on the Sustainability Outcomes of Spanish Wineries: An Exploration via Structural Equation Modeling

Introduction

Wine Tourism, Innovation, and Sustainable Winegrowing in Cool Climate Regions: A Longitudinal International Comparative Analysis

Tim Baird, C. Michael Hall, Pavel Castka, and Haywantee Ramkissoon

Introduction

Innovation within the context of New Zealand wine tourism represents an area which could potentially breathe new life into the local wine industry as it faces turbulent economic times (New Zealand Winegrowers, 2022a). However, as this study is based upon the OECD et al. (2005) definition of innovation, this suggests a potential dilemma for New Zealand wine tourism. On the one hand, you have the approach taken within the SWNZ scheme whereby the sustainable processes methods and products produced must meet a pre-determined criteria dictated by this organization (New Zealand Winegrowers, 2022b), while

T. Baird (✉)
Lincoln University, Christchurch, New Zealand
e-mail: tim.baird@lincoln.ac.nz

C. M. Hall
Department of Management, Marketing & Entrepreneurship, University of Canterbury, Christchurch, New Zealand
e-mail: michael.hall@canterbury.ac.nz

J. Martínez-Falcó et al. (eds.), *Wine Tourism and Sustainability*,
https://doi.org/10.1007/978-3-031-48937-2_1

on the other hand there are many wineries which pre-date the introduction of SWNZ in 1995 (New Zealand Winegrowers, 2022b) and arguably have already created their own innovations within the context of this definition without external intervention. Tourism and innovation policies in New Zealand appear to lack a significant form of interrelationship as noted by Hall (2009, p. 15) who points out that "it is possible that one of the reasons for the lack of recognition of tourism in innovation policy is that it is perceived as an industry that is not particularly innovative." Where there was a connection between tourism and innovation in New Zealand, strong institutional support on a local level has been seen to be a critical factor in enabling this connection (Hall, 2009).

This study examines wineries' perceptions towards innovation within the context of sustainable winegrowing practices and wine tourism in New Zealand. In order to investigate and contextualize the current situation in New Zealand, the National Wineries' Survey was revisited and extended to provide a longitudinal time series of data from the New Zealand wine industry. This data was also utilized to provide a comparison of the New Zealand perceptions of innovation with that of the Australian cool climate regions of Tasmania and Western Australia. It is important to note that Tasmanian and Western Australian wineries who take part in sustainable winegrowing programs are doing so voluntarily. This is not the case for their New Zealand counterparts, however; due to the mandatory membership required, New Zealand wine producers are required to be part of the Sustainable Wineries New Zealand (SWNZ) program (New Zealand Winegrowers, 2022b).

P. Castka
UC Business School, University of Canterbury, Christchurch, New Zealand
e-mail: pavel.castka@canterbury.ac.nz

H. Ramkissoon
UniSA International—University of South Australia, Adelaide, SA, Australia
e-mail: haywantee.ramkissoon@unisa.edu.au

Measuring Innovation Within the New Zealand Wine Industry Context

The New Zealand wine industry is at a formative point when it comes to research into the relationship between innovation, sustainable wine practices, and wine tourism. Very few prior studies have measured New Zealand wine industry responses to innovation; Cullen et al. (2013) are the exception through their research into the Greening Waipara scheme which is based in the Canterbury region of New Zealand. However, the 32 New Zealand vineyards involved in this particular study were not examined through the application of the OECD et al. (2005) framework that was used in this study. Applying innovation measures as set out by the OECD et al. (2005) to the New Zealand wine industry provides an opportunity to benchmark attitudes towards innovation within the context of past innovation studies conducted within the New Zealand agricultural sector. Innovations can signal significant changes to existing strategies, structures, and routines (Statistics New Zealand, 2013) and as such serve to grow the existing knowledge available for those involved in the supply-side of the industry (Cullen, et al., 2013; Hall, 2009; Hall & Williams, 2020).

Method

This cross-national survey utilized primary data obtained from participants who represent each of the 558 winegrowers located within New Zealand (Winetitles, 2015), the 125 winegrowers located in Tasmania (Winetitles, 2015), and 241 winegrowers located in Western Australia (Winetitles, 2015). These wineries all received copy of the survey via post, producing a combined total cross-national sample size of 924 wineries. A link to an online version of the survey was also provided to all participants should they prefer this particular option, and only two wineries from Western Australia elected to undertake this option. No electronic survey responses were received from either New Zealand or Tasmanian wineries.

Out of the 558 surveys that were sent out for the 2015 New Zealand survey, 145 surveys were received back. Of these 145 wineries, 80 responded that they had in fact recently gone out of business since the publication of *The 2015 Australian and New Zealand Wine Industry Directory* (Winetitles, 2015) but as these were still judged as valid responses, these surveys were still included in the overall response rate. The remaining 65 wineries who responded then provided the data that this study was based upon. This yielded a New Zealand response rate of 25.9%.

Innovation in Tasmanian, Western Australian, and New Zealand Wineries

The following section illustrates the levels of innovation within the Tasmanian and Western Australian wine industries in 2016 and compares these findings to the 2015 New Zealand survey results.

Introduction of Innovation

Figure 1 shows that New Zealand had the highest percentage when considering the introduction of new or improved good and services (46.2%) and new or significantly improved organizational processes (43.1%). Tasmania reported the highest level of implementation in terms of sales and marketing methods (53.3%) and new or significantly improved operational processes (46.7%). Western Australian wineries appeared to be the most risk averse when considering innovation; their percentages reported were the lowest out all three groups in all but one of the categories presented. This category (new or improved goods and services) was reported at 36.4% for Western Australia, which put them ahead of their Tasmanian counterparts (26.7%) but still behind New Zealand (46.2%).

Table 1 shows that based Tasmanian, Western Australian, and New Zealand winery size that none of the four categories were significant at

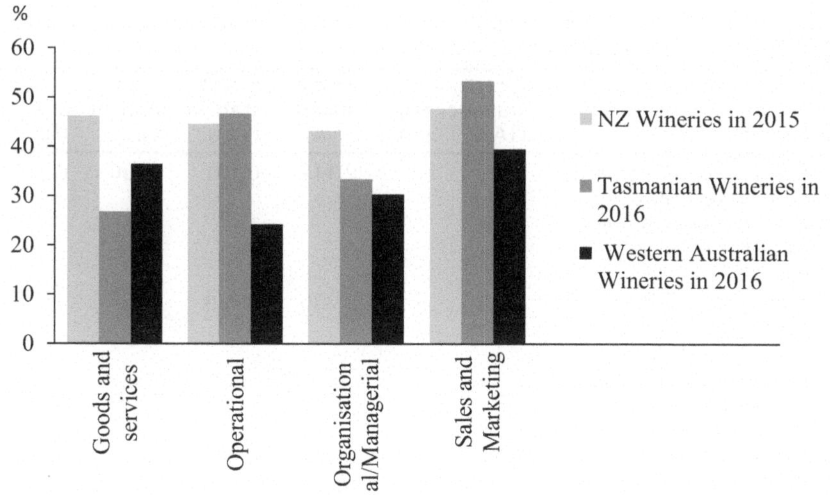

Fig. 1 Introduction of innovation over the two financial years prior to 2015/2016 for Tasmanian, Western Australian, and New Zealand wineries

the 0.05 level for Tasmanian wineries. The introduction of new or significantly improved operational processes where significant for both Western Australian (F = 3.235, Sig. = 0.24) and New Zealand wineries (F = 2.221, Sig. = 0.020). The other significant factor based on the size of the winery for Western Australian participants was the implementation of new or significantly improved organizational or managerial processes (F = 3.514, Sig. = 0.018). All other factors remained insignificant at the 0.05 level for Western Australian and New Zealand wineries.

Reasons for Innovation at Tasmanian, Western Australian, and New Zealand Wineries

A number of reasons were given for innovation at Tasmanian and Western Australian wineries (Table 2), with improving and increasing productivity being given as the primary reasons by Tasmanian wineries (both first equal at 80%). Reducing costs, increasing responsiveness to others, and establishing new market opportunities were second equal at 66.7% for Tasmanian wineries.

Table 1 Introduction of innovation over the two financial years prior to 2015/2016 based on winery size

Statement	*F* test statistic (TAS)	Sig. (TAS)	*F* test statistic (WA)	*F* test statistic (WA)	*F* test statistic (NZ)	Sig. (NZ)
Did your winery introduce any new or improved goods or services over the last 2 financial years?	–	–	2.111	0.101	1.136	0.377
Did your winery introduce on to the market any new or significantly improved operational processes over the last 2 financial years?	–	–	3.235	0.024	2.221	0.020
Did your winery implement any new or significantly improved organizational or managerial processes over the last 2 financial years?	0.497	0.820	3.514	0.018	1.039	0.470
Did your winery implement any new or significantly improved sales or marketing methods over the last 2 financial years?	–	–	1.476	0.256	0.936	0.583

Significance is measured at the 0.05 level
n = 15 (Tasmania), n = 33 (Western Australia), n = 65 (New Zealand)

In a reversal of this pattern, Western Australian wineries reported that reducing costs and increasing responsiveness to others were the most important reasons at 60.6% and 63.6%, respectively, while improving productivity was ranked second (45.5%) and increasing productivity and reducing environmental impact were in third place (42.4%). Establishing new market opportunities ranked much lower as a reason than their Tasmanian counterparts and was at 36.4%. The only Australian national and agricultural averages available from the ABS (2018) data were in the categories of increasing customer responsiveness (National average = 35.5%, Agricultural average = 11.6%), increasing market share (National average = 20.8%, Agricultural average = 18.7%), and

Table 2 Reasons for innovation occurring in Tasmanian, Western Australian, and New Zealand wineries

Reason	Yes (TAS) (%)	Yes (WA) (%)	Yes (NZ) (%)
To improve productivity	80.0	45.5	69.2
To increase productivity	80.0	42.4	50.8
To reduce costs	66.7	60.6	64.6
To increase responsiveness to customers	66.7	63.6	63.1
To increase market share	53.3	39.4	61.5
To establish/or exploit new market opportunities	66.7	36.4	63.1
To improve work safety standards	60.0	30.3	44.6
To reduce energy consumption	66.7	36.4	40.0
To reduce environmental impact	93.3	42.4	52.3
To replace goods and services being phased out	13.3	9.1	18.5

n = 15 (Tasmania), n = 33 (Western Australia), n = 65 (New Zealand)

establishing new market opportunities responsiveness (National average = 24.6%, Agricultural average = 19.7%). In each of these categories, all Tasmanian and Western Australian wineries reported figures that were above these national and agricultural averages.

New Zealand wineries gave improving productivity as their primary reason for innovation (69.2%). This was still below the 2013 New Zealand innovation average of 83% (Statistics New Zealand, 2013) and the 2013 New Zealand agricultural sector innovation average of 77% (Statistics New Zealand, 2013). The need to reduce costs ranked second (64.6%), followed by increasing responsiveness to customers and establishing or exploiting new market opportunities which were third equal at 63.1%. Other notable mentions were to increase market share (61.5%) and to reduce environmental impact (52.3%).

Table 3 details the reasons for innovation based on winery size for Tasmanian, Western Australian, and New Zealand wineries. No reasons are found to be significant for Tasmanian or New Zealand wineries at the 0.05 level of significance. However, for Western Australian wineries improving productivity (F = 3.762, Sig. = 0.014) was a significant reason for the occurrence of innovation based on the size of the winery concerned.

Table 3 Reasons for innovation occurring in Tasmanian, Western Australian, and New Zealand wineries based on winery size

Reason	*F* (TAS)	Sig. (TAS)	*F* (WA)	Sig. (WA)	*F* (NZ)	Sig. (NZ)
To improve productivity	–	–	3.762	0.014	0.744	0.800
To increase productivity	–	–	1.864	0.143	0.947	0.571
To reduce costs	–	–	1.184	0.398	0.655	0.884
To increase responsiveness to customers	–	–	0.739	0.735	0.870	0.660
To increase market share	–	–	1.369	0.301	0.712	0.833
To establish/or exploit new market opportunities	0.785	0.721	1.106	0.448	0.959	0.557
To improve work safety standards	–	–	1.812	0.155	0.701	0.843
To reduce energy consumption	–	–	1.310	0.330	1.160	0.356
To reduce environmental impact	–	–	1.679	0.188	0.984	0.530
To replace goods and services being phased out	0.682	0.752	1.038	0.494	1.088	0.422

Significance is measured at the 0.05 level
n = 15 (Tasmania), n = 33 (Western Australia), n = 65 (New Zealand)

Sources for Ideas and Information for Innovation at Wineries

When asked who or what were the main sources for ideas and information regarding innovation (Table 4) Tasmanian wineries cited books, journals patent disclosures, or the Internet as the leading source (86.7%), followed by competitors and other businesses from the wine industry (73.3%), and existing staff (66.7%). Government Research Institutes and Universities or polytechnics rated as the lowest sources of information (6.7%). Western Australian wineries regarded existing staff as the most valuable source for ideas and information (45.5%), followed by competitors and other businesses from the wine industry (42.4%). Universities or polytechnics rated as the least used source of information at 9.1%.

Leading sources of ideas and information for innovation at New Zealand wineries were existing staff (62.4%), new staff (46.2%), and competitors and other businesses within the wine industry (44.6%). From the data that was available for the Australian national and agricultural averages, it was found that customers (National average = 21.5%, Agricultural average = 38.6%), suppliers (National average = 25.2%, Agricultural average = 21.8%) along with competitors, and other businesses within the wine industry (National average = 24%, Agricultural average = 27.9%) were all categories where wineries were above both the national and agricultural averages. The category of industry or employer organizations was found to be where the combined Australian wineries total of 33.3% was below the national average of 36.6% but above the agricultural average of 15.5%.

Sources for Ideas and Information for Innovation at Wineries Based on Winery Size

Table 5 shows that no sources for ideas and information for innovation were significant at the 0.05 level for Tasmanian or New Zealand wineries. Western Australian wineries, however, reported that competitors and other businesses within the wine industry ($F = 3.912$, Sig. = 0.012) and the use of books, journals patent disclosures, or the Internet ($F = 3.014$, Sig. = 0.031) were significant at the 0.05 level. Government Research Institutes, other research institutes, or research associations were also found to be significant ($F = 5.254$, Sig. = 0.003) as well.

Table 4 Sources of ideas and information for innovation at Australian and New Zealand wineries

Source	Yes (TAS) (%)	Yes (WA) (%)	Yes (AU Comb.) (%)	AU national average (2016) (%)	AU agricultural average (2016) (%)	Yes (NZ) (%)	NZ national average (2013) (%)	NZ agricultural average (2013) (%)
New staff (appointed in the last 2 years)	40.0	30.3	40.0	–	–	46.2	52.0	48.0
Existing staff	66.7	45.5	66.7	–	–	64.6	73.0	71.0
Other businesses within the business group	20.0	15.2	20.0	49.5	61.1	21.5	47.0	–
Customers	53.3	39.4	53.3	21.5	38.6	46.2	61.0	26.0
Suppliers	53.3	30.3	53.3	25.2	21.8	40.0	50.0	40.0
Competitors and other businesses from the same industry	73.3	42.4	73.3	24.0	27.9	44.6	N/A	–
Businesses from other industries (not including customers or suppliers)	20.0	27.3	20.0	–	–	26.2	N/A	–
Consultants, banks, or accountants	26.7	24.2	26.7	–	–	33.8	49.0	58.0
Books, journals, patents, or the Internet	80.0	39.4	80.0	–	–	38.5	46.0	47.0
Wine shows, festivals, or conferences	60.0	33.3	60.0	–	–	30.8	43.0	47.0
Industry or employer organizations	33.3	15.2	33.3	36.6	15.5	30.8	31.0	38.0
Universities or polytechnics	6.7	9.1	6.7	4.8	3.0	6.2	10.0	11.0
Government Research Institutes, other research institutes, or research associations			6.7	3.0	1.0	12.3	12.3	15.0
Government agencies			20.0	–	2.7	7.7	7.7	5.0

$n = 15$ (Tasmania), $n = 33$ (Western Australia), $n = 48$ (AU Combined), $n = 65$ (New Zealand)

Table 5 Sources of ideas and information for innovation at Tasmanian, Western Australian, and New Zealand wineries based on winery size

Source	*F* (TAS)	Sig (TAS)	*F* (WA)	Sig. (WA)	*F* (NZ)	Sig. (NZ)
New staff (appointed in the last 2 years)	0.744	0.733	1.562	0.225	0.986	0.528
Existing staff	–	–	1.250	0.361	1.154	0.361
Other businesses within the business group (e.g., subsidiaries or parent companies)	–	–	0.746	0.729	1.093	0.417
Customers	0.497	0.820	0.891	0.607	0.768	0.774
Suppliers	–	–	2.460	0.063	0.966	0.550
Competitors and other businesses from the same industry	–	–	3.912	0.012	1.446	0.170
Businesses from other industries (not including customers or suppliers)	–	–	0.683	0.783	1.683	0.089
Professional advisors, consultants, banks, or accountants	–	–	1.290	0.339	0.683	0.860
Books, journals, patent disclosures, or the Internet	–	–	3.014	0.031	0.513	0.970
Wine shows, festivals, or conferences	0.477	0.829	2.037	0.112	1.135	0.378
Industry or employer organizations	0.682	0.752	1.127	0.434	0.604	0.923
Universities, TAFE Colleges, or polytechnics	0.231	0.942	1.102	0.450	0.740	0.805
Crown/Government Research Institutes, other research institutes, or research associations	–	–	5.254	0.003	0.670	0.871
Government agencies	–	–	2.439	0.065	0.477	0.981

Significance is measured at the 0.05 level
$n = 15$ (Tasmania), $n = 33$ (Western Australia), $n = 65$ (New Zealand)

Activities to Support Innovation at Tasmanian, Western Australian, and New Zealand Wineries

Table 6 details the activities that were undertaken to support innovation at Tasmanian and Western Australian wineries. The main activities listed by Tasmanian wineries that were done in support of innovation were the implementation of organizational restructuring and acquiring new machinery and equipment (both 60%). Design-related activities and employee training were both undertaken by 53.3% of respondents. Western Australian wineries listed marketing the introduction of new goods and services as the activity most likely to be performed in to support innovation (27.3%). The use of new viticultural techniques and design were both undertaken by 24.2% of respondents. No data for the Australian national or agricultural averages was available.

When compared with activities that New Zealand wineries performed to support innovation the main activities listed were design at 36.9%, followed by the acquisition of new computer hardware or software (32.3%) and the acquisition of new machinery or equipment (30.8%). Table 7 shows that at the 0.05 level of significance that no activities were significant for Tasmanian wineries based on winery size. Western Australian wineries listed acquiring other knowledge ($F = 3.000$, Sig. = 0.032), implementing new business strategies ($F = 3.222$, Sig. = 0.025), and implementing organizational restructuring ($F = 2.680$, Sig. = 0.047) as significant. Making significant changes to existing marketing strategies was also listed as being significant ($F = 10.048$, Sig. = 000) for Western Australian wineries. The only activity listed as significant at the 0.05 level for New Zealand wineries was the acquisition of new machinery and equipment ($F = 2.346$, Sig. = 0.015).

Table 6 Activities to support innovation at Tasmanian Western Australian and New Zealand wineries

Activity	Done to support innovation (TAS) (%)	Done, though not to support innovation (TAS) (%)	Not Done (TAS) (%)	Don't know (TAS) (%)	Done to support innovation (WA) (%)	Done, though not to support innovation (WA) (%)
Introduce a new variety of grape	26.7	73.3	0.0	0.0	18.2	6.1
Use of new viticultural techniques	40.0	27.7	33.3	0.0	24.2	21.2
Acquire new machinery and equipment	60.0	26.7	13.3	0.0	21.2	30.3
Acquire new computer hardware and software	33.0	20.0	46.7	0.0	6.1	36.4
Acquire other knowledge (e.g., outsourcing for new techniques or intellectual property)	26.7	46.7	26.7	0.0	18.2	9.1
Implement new business strategies or management techniques	33.3	27.7	40.0	0.0	15.2	9.1
Implement organizational restructuring	60.0	6.7	33.3	0.0	18.2	6.1
Design (e.g., graphic design on labeling of bottles)	53.3	6.7	40.0	0.0	24.2	15.2
Market the introduction of new goods and services	26.7	26.7	40.0	0.0	27.3	12.1
Market research	27.6	40.0	33.3	0.0	12.1	12.1

(continued)

Table 6 (continued)

Activity	Done to support innovation (TAS) (%)	Done, though not to support innovation (TAS) (%)	Not Done (TAS) (%)	Don't know (TAS) (%)	Done to support innovation (WA) (%)	Done, though not to support innovation (WA) (%)
Significant changes to existing marketing strategies	33.3	20.0	46.7	0.0	18.2	18.2
Employee training	53.3	13.3	33.3	0.0	12.1	15.2

Activity	Not Done (WA) (%)	Don't know (WA) (%)	Done to support innovation (NZ) (%)	Done, though not to support innovation (NZ) (%)	Not Done (NZ) (%)	Don't know (NZ) (%)
Introduce a new variety of grape	75.8	0.0	29.2	6.2	60.0	4.6
Use of new viticultural techniques	54.5	0.0	23.1	20.0	52.3	4.6
Acquire new machinery and equipment	48.5	0.0	30.8	23.1	44.6	1.5
Acquire new computer hardware and software	57.6	0.0	32.3	24.6	41.5	1.5
Acquire other knowledge (e.g., outsourcing for new techniques or intellectual property)	72.7	0.0	21.5	16.9	56.9	4.6
Implement new business strategies or management techniques	75.8	0.0	26.2	18.5	52.3	3.1

Activity	Not Done (WA) (%)	Don't know (WA) (%)	Done to support innovation (NZ) (%)	Done, though not to support innovation (NZ) (%)	Not Done (NZ) (%)	Don't know (NZ) (%)
Implement organizational restructuring	57.6	0.0	24.6	30.8	40.0	4.6
Design (e.g., graphic design on labeling of bottles)	60.6	0.0	36.9	21.5	38.5	3.1
Market the introduction of new goods and services	60.6	0.0	29.2	16.9	52.3	1.5
Market research	75.8	0.0	26.2	18.5	53.8	1.5
Significant changes to existing marketing strategies	63.6	0.0	26.2	24.6	47.7	1.5
Employee training	72.7	0.0	29.2	33.8	33.8	3.1

$n = 15$ (Tasmania), $n = 33$ (Western Australia), $n = 65$ (New Zealand)

Table 7 Activities to support innovation at Tasmanian, Western Australian, and New Zealand wineries based on winery size

Activity	*F* (TAS)	Sig. (TAS)	*F* (WA)	Sig. (WA)	*F* (NZ)	Sig. (NZ)
Introduce a new variety of grape	0.374	0.874	0.877	0.619	1.578	0.119
Use of new viticultural techniques	1.605	0.556	1.098	0.453	1.241	0.291
Acquire new machinery and equipment	–	–	1.868	0.143	2.346	0.015
Acquire new computer hardware and software	0.374	0.874	–	–	1.293	0.255
Acquire other knowledge (e.g., outsourcing for new techniques or intellectual property)	0.231	0.942	3.000	0.032	1.239	0.293
Implement new business strategies or management techniques	–	–	3.222	0.025	1.721	0.080
Implement organizational restructuring	–	–	2.680	0.047	0.974	0.541
Design (e.g., graphic design on labeling of bottles)	0.451	0.840	1.245	0.363	1.323	0.236
Market the introduction of new goods and services	0.323	0.898	1.628	0.203	1.060	0.450
Market research	–	–	0.988	0.531	0.974	0.541
Significant changes to existing marketing strategies	–	–	10.048	0.000	1.139	0.374
Employee training	–	–	2.065	0.107	0.891	0.635

Significance is measured at the 0.05 level. $n = 15$ (Tasmania), $n = 33$ (Western Australia), $n = 65$ (New Zealand)

Sustainability in Tasmanian, Western Australian, and New Zealand Wineries

Sustainable Winegrowing Program Certification for Tasmanian and Western Australian Wineries

Only 40% of Tasmanian respondents stated that they were currently a member of a sustainable winegrowing program. In terms of the answers given to this question there appeared to be some confusion among the remaining 60% of respondents as to whether there was in fact a locally based sustainable winegrowing program that they could be a member of. This pointed towards possible problems in the communication between the wineries concerned and those who oversee developing and implementing these programs in Tasmania. Only 21.2% of their Western Australian counterparts indicated that they were currently members of a certified sustainable winegrowing program. Both the individual Western Australian and Tasmanian membership figures fall well below the 78.5% SWNZ membership figure reported by New Zealand wineries in the 2015 survey.

Level of Commitment Towards Sustainability

Wineries were given the opportunity to demonstrate their commitment towards sustainable practices (Table 8). Participants were asked to state whether they were interested in adopting sustainable practices or were planning to adopt sustainable practices or had already adopted such practices. An option was also provided for those who were not interested in including sustainable practices as part of their business operations.

When asked about the level of commitment that their winery had towards sustainability, 56.3% of Western Australian wineries specified that they had adopted sustainable practices; 6.3% were planning to adopt sustainable practices, while a further 25% confirmed that they were interested in considering adopting these practices. 12.5% of Western Australian wineries asserted that they were not interested in the pursuit of sustainable practices.

Table 8 Level of commitment towards sustainability among Tasmanian, Western Australian, and New Zealand wineries

	Not interested in adopting sustainable practices (%)	Interested in adopting sustainable practices (%)	Planning to adopt sustainable practices but not ready yet (%)	Have already adopted sustainable practices (%)
Tasmania	0.0	0.0	13.3	86.7
Western Australia	12.5	25.0	6.3	56.3
New Zealand	3.1	3.1	0.0	93.8

n = 15 (Tasmania), n = 33 (Western Australia), n = 65 (New Zealand)

86.7% of Tasmanian wineries surveyed stated that they had already adopted sustainable winegrowing practices, while 13.3% reported that they were planning on incorporating sustainability in their business practices. No wineries indicated that they were not interested in sustainability. This demonstrates that Tasmanian wineries have differing attitudes towards whether commitment towards sustainability also entails sustainable winegrowing program membership.

Importance of Sustainable Practices on Attracting Winery Visitation

Western Australian wineries held differing perspectives on whether sustainable practices helped to attract winery visitation (Table 9). 31.3% of participants reported that sustainable practices were important for attracting wine tourists, while 28.8% indicated that this was somewhat important to their business operations. 18.8% felt that attracting wine tourists was very important, while 15.6% stated that this was not important whatsoever. The remaining 6.3% reported that sustainable business practices were extremely important in terms of attracting visitors to their wineries.

The impact of sustainable practices on wine tourism was also noted by Tasmanian wineries with 40% of respondents declaring that they felt this was important for their business. 26.7% respectively stated that they

Table 9 Importance of sustainable practices for Tasmanian, Western Australian, and New Zealand wineries

	TAS					
Sustainable practice	Not important (%)	Somewhat important (%)	Important (%)	Very important (%)	Extremely important (%)	Mean
Stronger brand/ greater pricing power	0.0	6.7	40.0	33.3	20.0	3.67
Employee recruitment, morale and retention	0.0	26.7	46.7	20.0	6.7	3.07
Greater operational efficiency	0.0	0.0	26.7	53.3	20.0	3.93
Improved customer loyalty	0.0	6.7	73.3	20,0	0.0	3.13
Enhanced ability to enter new markets	0.0	0.0	33.3	46.7	20.0	3.87
Migrant workers' rights	0.0	0.0	20.0	53.3	26.7	4.07
More potential sources of revenue	0.0	6.7	46.7	33.3	13.3	3.53
Lower financial and operating risk	0.0	20.0	46.7	20.0	13.3	3.27
Ability to justify price premium for products	0.0	0.0	33.3	46.7	20.0	3.87
Attracting visitors to winery	6.7	26.7	40.0	26.7	0.0	2.87
Lower legal and regulatory risk	0.0	6.7	60.0	13.3	20.0	3.47

(continued)

Table 9 (continued)

	WA					
Sustainable practice	Not important (%)	Somewhat important (%)	Important (%)	Very important (%)	Extremely important (%)	Mean
Stronger brand/ greater pricing power	12.5	15.6	34.4	28.1	9.4	3.09
Employee recruitment, morale and retention	9.4	40.6	25.0	25.0	15.6	3.27
Greater operational efficiency	12.5	18.8	43.8	18.8	6.3	2.88
Improved customer loyalty	12.5	12.5	21.9	21.9	31.3	3.52
Enhanced ability to enter new markets	6.3	6.3	28.1	25.0	34.4	3.79
Migrant workers' rights	6.3	9.4	34.4	43.8	6.3	3.36
More potential sources of revenue	12.5	18.8	37.5	21.9	9.4	2.97
Lower financial and operating risk	15.6	28.1	43.8	3.1	9.4	2.61
Ability to justify price premium for products	6.3	15.6	40.6	21.9	15.6	3.24
Attracting visitors to winery	15.6	28.1	31.3	18.8	6.3	2.70
Lower legal and regulatory risk	6.3	12.5	31.3	25.0	25.0	3.48

	NZ					
Sustainable practice	Not important (%)	Somewhat important (%)	Important (%)	Very important (%)	Extremely important (%)	Mean
Stronger brand/ greater pricing power	29.2	23.1	16.9	13.8	16.9	2.66
Employee recruitment, morale and retention	27.7	13.8	24.6	18.5	15.4	2.80
Greater operational efficiency	23.1	16.9	23.1	24.6	12.3	2.86
Improved customer loyalty	27.7	12.3	30.8	16.9	12.3	2.74
Enhanced ability to enter new markets	29.2	7.7	30.8	20.0	12.3	2.78
Migrant workers' rights	47.7	13.8	26.2	4.6	7.7	2.11
More potential sources of revenue	35.4	21.5	27.7	3.1	12.3	2.35
Lower financial and operating risk	43.1	15.4	21.5	9.2	10.8	2.29
Ability to justify price premium for products	23.1	18.5	36.9	7.7	13.8	2.71
Attracting visitors to winery	30.8	16.9	36.9	7.7	7.7	2.44
Lower legal and regulatory risk	27.7	20.0	23.1	18.5	10.8	2.65

$n = 15$ (Tasmania), $n = 33$ (Western Australia), $n = 65$ (New Zealand)

believed that attracting wine tourists through engaging in sustainable practices was either very important or somewhat important. Only 6.7% of Tasmanian wineries surveyed indicated that they felt that this area was of no importance.

New Zealand wineries appeared to have a spectrum of differing viewpoints when considering this topic. 36.9% of respondents indicated that they felt that sustainable practices were important in this regard; however, 30.8% also felt that sustainable practices were not all important in terms of attracting wine tourists, which was the highest figure for this category over the three survey groups.

Discussion

Introduction of Innovation

When considering the degree of innovativeness exhibited by all three survey groups it was apparent that there were elements of risk aversion present. However, this aversion existed within different categories for each group. Risk aversion among wineries towards innovation is not a new phenomenon; this pattern had also presented itself during the 2010 New Zealand wineries survey where the implementation of innovation within existing goods and services was noted as one particular area which local wineries were reluctant to pursue (Baird, 2012). The findings in this study have found that this risk aversion still exists, and not only at a local but at a cross-national level as well.

The lack of sales revenue which was directly attributed to goods or services which had undergone significant improvement further demonstrates that all three survey groups appeared to be approaching innovation with a degree of caution. The wine industries within which each of the three survey groups operate are not reliant on the constant introduction of innovation in order to assure ongoing growth in their existing revenue streams. For example, New Zealand wineries decreed that less than 10% of sales came from significantly improved goods or services, indicating that the main forms of innovation being undertaken at the time this survey was undertaken lie within the other categories that were being investigated.

Improvements Made to Business Processes

The desire to employ innovative practices to improve operational processes was one area where there was evidence of a slightly more proactive approach on the part of Tasmanian and New Zealand wineries. Wineries were able to make these improvements either by themselves or in partnership with others. The results of the 2010 New Zealand survey indicated that there is a gradual move away from past reluctance to implement or adopt new or significantly improved operational processes. However, major questions remain with respect to the differences in cooperative practices and their differential use for innovation, i.e., cooperation may occur on innovation in some areas but not others, but this varies between regions. For example, Western Australian wineries did not have a focus on continuous improvement of their operational processes. Therefore, the question remains as to whether the different focus on innovation within networks is a function of regional needs, the activities of key actors, or some combination of all three factors.

Similar issues occurred with respect to improvements made to organizational and managerial processes. For example, Tasmanian wineries placed great emphasis on partnership in relation to improvements made to organizational and managerial processes which was well above the Australian national average. This supports past studies which have highlighted the significance of process innovations within the wine industry (Baird, 2012; Bessant et al., 2009; Doloreux et al., 2013; Ohmart, 2008; Yuan et al., 2006). New Zealand wineries also demonstrated levels of innovation in this area that were well above the national and agricultural averages. However, Western Australian wineries were well behind national levels of innovation in this area.

Having the ability to make improvements to sales and marketing methods proved to be the most important area for innovation across all three survey groups. Not only was in-house development of these methods occurring, but also the importance of knowledge sharing and collaboration was noted, particularly for Tasmanian wineries (refer to Table 6). This reflects past international research into this area (Gilinsky et al., 2008) and also reiterates the importance within the wine industry of networks that enable the dissemination of knowledge in order to both increase revenue and create brand awareness for local wineries via improved marketing (Alonso & Bressan, 2013; Doloreux et al., 2013; Pickering et al., 2012; Taplin & Breckenridge, 2008). What is currently unknown, however, is whether innovation in sales and marketing is purely related to winery level marketing or also includes support for regional wine marketing initiatives. Similarly, there is no research in the wine marketing innovation literature as to whether changed practices are domestically or internationally oriented.

Reasons for Innovation

A wide range of reasons was given for innovation across the three survey groups. Nevertheless, the desire to become more productive while simultaneously finding ways in which to reduce costs was a common theme. Despite concerns over sustainable practices which have been previously discussed, both New Zealand and Western Australian wineries cited the need to reduce environmental impact as being one of their key reasons for innovation. Also notable was the value placed by survey participants across all three groups on being able to respond to new market opportunities through embracing innovation. This suggests that given the apparent initial risk aversion shown by wineries in terms of considering innovation when it came goods and services and their operational processes that a point may have been reached where innovation can no longer be ignored. This is particularly pertinent if there are distinct and clearly perceived economic and environmental benefits which can be derived from introducing new or improved techniques or technologies.

Sources of Ideas, Activities, and Information for Innovation

The ability of winery staff to provide information and new ideas which could serve to inform innovation was a common theme among all three survey groups. However, relying on staff was not the only way in which wineries sourced ideas; online information and books and journals also proved popular. A point of interest with regard to this topic is that the pattern noted in the 2010 survey where government research institutes were cited as being among the least used source for information regarding innovation also occurred in the New Zealand 2015 survey; this trend was also mirrored by Australian survey participants as well. This suggests that knowledge transfer in this area is potentially poor despite the development of wine and viticultural research clusters between industry and institutions.

The range of activities that were cited by the three survey groups illustrates that each group is at a different phase in terms of what is valued as being appropriate in terms of supporting mechanisms to aid innovative activity. A point to note in this regard is that apart from the acquisition of new machinery to improve viticultural practices, aspects of marketing such as design were high on the agendas of many survey participants. The value placed on product knowledge and service at not only the cellar door but at the production level as well is reflected by employee training and the acquisition of knowledge both being ranked highly as innovative activities. This was particularly pertinent for Western Australian wineries where the size of the winery (refer to Table 7) was found to be a significant factor in the undertaking of such activities. This finding supports the research of Spielmann (2017) who recognized that winery size was a determinant on whether or not wineries adopted innovation. Running an efficient wine tourist attraction using innovative techniques was the focus of a study by Karagiannis and Metaxas (2019); findings from this research also indicated that winery size not only determined whether or not wineries were able to implement innovations, but also if the wineries concerned were able to sustain these innovations for over a long-term period (Karagiannis & Metaxas, 2019).

Conclusion

In terms of environmental issues, leading reasons for the employment of sustainable methods were found to be the reduction of environmental impacts, the reduction of costs, and the improvement of productivity. Attracting winery visitors, and in turn also seeking to increase revenue through openly advocating sustainability, were not viewed by wineries as motivating factors to adopt these methods.

It appears that for cool climate wineries engagement in sustainable winegrowing techniques is not an automatic guarantee that innovation is taking place. In New Zealand, the pattern of risk aversion was first noticed in the 2010 National Wineries' Survey remained in the 2015 survey iteration. Wineries were only slightly more innovative than they had been in the previous survey; risk aversion still appeared to be present, and New Zealand wineries wanted to see a proven track record of success before adopting new viticultural techniques or processes. Western Australian wineries were found to be the most open to innovation, but even then, an air of caution still exists.

References

Alonso, A. D., & Bressan, A. (2013). Stakeholders' perspectives on the evolution and benefits of a traditional wine festival: The case of the Grape Festival ("Festa dell'Uva") in Impruneta, Italy. *Journal of Convention & Event Tourism, 14*(4), 309–330.

Australian Bureau of Statistics [ABS]. (2018). *Innovation in Australian Business, 2016–17*. https://www.abs.gov.au/AUSSTATS/abs@.nsf/DetailsPage/8158.02016-17?OpenDocument

Baird, T. (2012). *Between the vines: A comparative analysis of wineries' attitudes towards wine tourism in New Zealand* (Unpublished master's thesis). University of Canterbury, Christchurch, New Zealand.

Bessant, J., Tsekouras, G., & Rush, H. (2009, September 6–8). *Getting the tail to wag: Developing innovations capability in SMEs*. Paper presented at CI Net 2009, Brisbane, Australia.

Cullen, R., Forbes, S., & Grout, R. (2013). Non-adoption of environmental innovations in wine growing. *New Zealand Journal of Crop and Horticultural Science, 41*(10), 41–48.

Doloreux, D., Chamberlin, T., & Ben-Amor, S. (2013). Modes of innovation in the Canadian wine industry. *International Journal of Wine Business Research, 25*(1), 6–26.

Gilinsky, A., Santini, C., Lazzeretti, L., & Eyler, R. (2008). Desperately seeking serendipity: Exploring the impact of country location on innovation in the wine industry. *International Journal of Wine Business Research, 20*(4), 302–320.

Hall, C. M. (2009). Innovation and tourism policy in Australia and New Zealand: Never the twain shall meet? *Journal of Policy Research in Tourism Leisure and Events, 1*(1), 2–18.

Hall, C. M., & Williams, A. M. (2020). *Tourism and innovation*. Routledge.

Karagiannis, D., & Metaxas, T. (2019). Innovation in wine tourism businesses: 'Turning ashes to gold'. *Management and Marketing of Wine Tourism Business: Theory, Practice, and Cases*, 345–363.

New Zealand Winegrower.s (2022a). *2022 sustainability report*. https://www.nzwine.com/en/sustainability/sustainability-report/

New Zealand Winegrowers. (2022b). *Sustainable Winegrowing New Zealand*. https://www.nzwine.com/en/sustainability/sustainable-winegrowing-nz

OECD and Statistical Office of European Communities. (2005). *Oslo manual: Guidelines for collecting and interpreting innovation data* (3rd ed.). OECD.

Ohmart, C. (2008). Innovative outreach increases adoption of sustainable winegrowing practices in Lodi region. *California Agriculture, 62*(4), 142–147.

Pickering, K., Pickering, G., Inglis, D., Shaw, T., & Plummer, R. (2012, June 14–16). Innovation and adaptation in the Ontario grape and wine industry: An integrated transdisciplinary response to climate change. In *Proceedings of the International Conference on Innovation and Trends in Wine Management*. Frankfurt, Germany.

Spielmann, N. (2017). Larger and better: Examining how winery size and foreign investments interact with sustainability attitudes and practices. *International Journal of Wine Business Research, 29*(2), 178–194.

Statistics New Zealand. (2013). *Innovation in New Zealand: 2013*. http://archive.stats.govt.nz/browse_for_stats/businesses/business_growth_and_innovation/innovation-in-new-zealand-2013-tables.aspx#gsc.tab=0

Taplin, I. M., & Breckenridge, R. S. (2008). Large firms, legitimation and industry identity: The growth of the North Carolina wine industry. *The Social Science Journal, 45*, 352–360.

Winetitles. (2015). *The Australian and New Zealand wine industry directory*. Hartley-Higgins.

Yuan, Y. L., Gretzel, U., & Fesenmaier, D. R. (2006). The role of information technology use in American convention and visitor bureaus. *Tourism Management, 27*(3), 326–341.

Economic Impact of Wine Tourism

Economic Structure and Geographic Scope of Spanish Wine Routes

Luis A. Millán-Tudela, Bartolomé Marco-Lajara, Javier Martínez-Falcó, and Eduardo Sánchez-García

Introduction

Porter (1998) defines clusters (also called industrial districts) as "geographic concentrations of interconnected firms, specialized suppliers, service providers, firms in related industries, and associated institutions in particular fields in which there is competition, but also cooperation." Within this conception, Martínez-Falcó et al. (2023) identify wine

L. A. Millán-Tudela (✉) · B. Marco-Lajara · J. Martínez-Falcó · E. Sánchez-García
University of Alicante, Alicante, Spain
e-mail: luisantonio.millan@ua.es

B. Marco-Lajara
e-mail: bartolome.marco@ua.es

J. Martínez-Falcó
e-mail: javier.falco@ua.es

E. Sánchez-García
e-mail: eduardo.sanchez@ua.es

J. Martínez-Falcó et al. (eds.), *Wine Tourism and Sustainability*,
https://doi.org/10.1007/978-3-031-48937-2_2

tourism activities "structured as a cluster through wine routes." From these definitions, it can be deduced that wine routes constitute the physical environment in which the aforementioned agents are concentrated, in this case under the umbrella of wine tourism activity.

The existence of wine routes can be useful for various purposes. One of them can be the promotion of rural areas that are highly affected by depopulation (Marco-Lajara et al., 2023). Indeed, there are large regions that, despite having a multitude of exploitable natural resources, are in a situation of low population density as a result of their concentration in large urban centers. Even so, there are other reasons that justify the promotion of wine tourism from a strategic perspective. For example, the increase in knowledge achieved through this type of initiative can lead to a strengthening of the cultural elements of the various regions, and even to the promotion of biodiversity in these areas, since it allows the richness of the natural environments in which this economic activity takes place to be exposed.

From a more abstract point of view, although equally necessary, it should be considered that the particularities of each region give rise to a series of specific needs that the different organizations must meet in order to achieve congruence with their environment, which Guerras Martín and Navas López (2022) refer to as strategic adjustment. In fact, after achieving this adjustment, the companies, in this case the different wineries, are required to adapt so that the strategy in question is congruent with their own internal reality, which is called organizational adjustment. For this reason, the analysis of the different characteristics of wine routes is crucial. On the one hand, an aggregate analysis makes it possible to understand both the structure of each route and the differences between them. Moreover, since this structure arises as a consequence of the structure of the organizations that compose it, a better understanding of their characteristics is also achieved.

Given the above, the main objective of this chapter is to carry out a descriptive analysis of the structure of the different Spanish wine routes. More specifically, the potential geographical scope of each of them in the Iberian Peninsula is analyzed, as well as the longevity of the organizations that comprise them. In addition, an analysis of the main accounting figures of the wineries is carried out. With all this, a series of generic strategic actions are presented that seek to improve the generation of wealth in the different territories in which these routes are located, thus contributing to the development of public, private, and public–private plans that, in the medium and long term, will improve the economic sustainability of these areas.

For this purpose, the chapter is structured as follows. After the introduction, the methodology section is presented, detailing both the use of confidence intervals with prior data transformation and analysis of variance as tools for comparing the different routes. The so-called Voronoi diagram is also explained as a proxy for the geographic scope of each of the wine routes. After this, we proceed to present the results obtained with both tools, comparing the different routes on the basis of their book values and grouping them into clusters based on their geographical scope. Finally, the paper closes by discussing the implications of these results for the different routes, as well as the issues related to the particularities of the present research.

Methodology

As previously indicated, one of the main interests of the different wine routes is to publicize the particularities of their emplacements, since a greater exposure of these leads to a greater appreciation of the virtues that exist in them, which in turn can enhance the generation of wealth associated with these strengths. However, the exploitation of these strengths cannot be carried out indiscriminately, since the potential visitors are a determining factor in the strategy to be adopted. For this reason, an estimate must be made of the scope of each of the different routes in the Iberian Peninsula. For this purpose, the so-called Voronoi diagram is used, which is defined as the "regions of proximity for a series of positions in the plane where the distance of the points is defined by their Euclidean space" (Erwig, 2000). In essence, Voronoi diagram indicates the regions of space that are closest to a given point of interest with respect to other points of interest. Thus, knowing the locations of the wine routes, this technique makes it possible to establish areas of influence based on the closest existing point of interest. Obtaining it manually is extremely complex, so its use is based on the use of computer algorithms. In this case, the *deldir* software package developed by Turner (2021) in the R programming language is applied. For this purpose, the headquarters of each of the routes is set as the starting location, generating each of the areas from them. In this way, each wine route presents an area in which its headquarters is the closest of all the existing ones, which is used as a proxy for their potential geographical scope.

Following this, an analysis of both the longevity and the accounting structure of the wineries of the routes considered is carried out. To this end, a series of statistical tests are applied to the annual report data of the companies corresponding to the 2021 financial year. On the one hand, the Kruskall-Wallis test is applied for the analysis of variance, since various tests for the verification of the assumptions of normality and homogeneity of variance invalidate the application of the ANOVA technique (Goss-Sampson, 2022). Additionally, the overlap of the confidence intervals for the different variables considered is analyzed, so that a disaggregation of the population of routes based on their structural characteristics is achieved.

To develop all of the above, the SABI database (Bureau van Dijk, 2023) was used to collect all the information from the target population, which is composed of a total of 747 wineries ($N = 747$). By discarding those records with missing values, a sample of 402 wineries was obtained, which represents 53.8% of the population. Thus, the authors are confident about the representativeness of this population.

Results

Potential Geographical Scope

The application of the Voronoi diagram has been based on the positioning of a total of 33 wine routes in the Spanish peninsular territory, resulting in the distribution shown in Fig. 1.

As can be seen, there are a series of distribution patterns that should be highlighted. On the one hand, there is great atomization in the region linked to the area of La Rioja, which in turn constitutes one

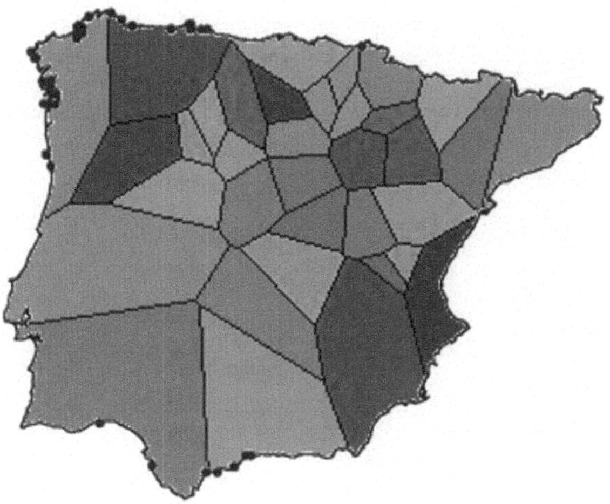

Fig. 1 Voronoi diagram in the Iberian Peninsula for the different Spanish wine routes (*Source* Elaborated from Turner [2021])

of the main strongholds of Spanish appellations of origin in terms of winemaking. There is also great potential in the southern area of the peninsula. With the exception of the Rías Baixas and Marco de Jerez wine routes, which could present a greater scope as a consequence of the absence of Portuguese routes in this analysis, the rest of the routes in the Atlantic region, as well as in the southern Mediterranean, show large areas of influence. In addition, it is possible to defend an average scope in the rest of the regions, essentially in the east and northeast of the peninsula, as well as in the areas of Lugo (Galicia) and Asturias, as well as in the area of León. In essence, a compartmentalization by groups can be made on the basis of their geographic scope. This division is shown in Fig. 2, where those routes with high reach are shown in dotted orange pattern, medium reach is shown in dashed blue, while the group with the highest atomization is shown in green grid.

Average Longevity of Wine Routes' Wineries

To determine the average longevity of the wineries on each route, the year of establishment of each of the organizations was used. For this

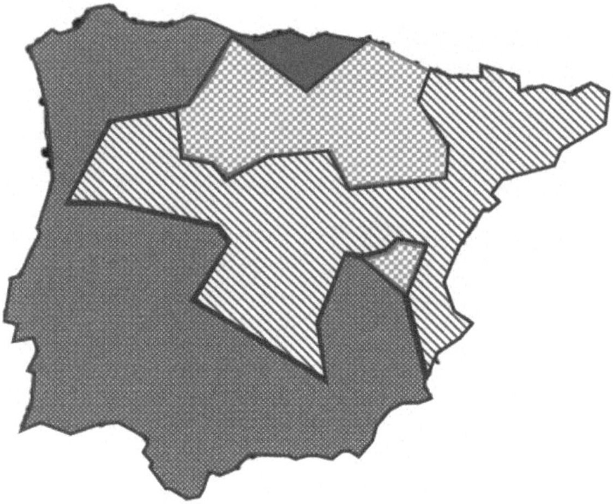

Fig. 2 Sorting of wine routes based on their geographical scope (*Source* Own elaboration)

purpose, the records in the SABI database (Bureau van Dijk, 2023) were used. Based on the available results, the different wineries were grouped according to the route or routes to which they are linked, obtaining the average value for each of them. Although the initial objective was to carry out an analysis of variance using the ANOVA technique, the verification of the assumptions of normality and homogeneity of variance has yielded results that demonstrate non-compliance. Therefore, and following the recommendations of Goss-Sampson (2022), the Kruskall-Wallis test was resorted, which is a non-parametric alternative to the ANOVA technique. After this, the data was transformed to produce confidence intervals so that, as with the geographic range, groups of routes with statistically similar longevity values could be shown. Table 1 shows the results of the Kruskall-Wallis test. In this case, the number of routes previously considered has been reduced to 21, since the number of wineries linked to some them is less than 30, which can generate problems in the application of the contrast techniques, so they have been removed from the analysis.

As can be seen, statistically significant differences are obtained at the level $\alpha = 0.01$, showing that at least two routes present statistically

Table 1 Kruskall-Wallis test for longevity of wine routes' wineries

Variable	Kruskall-Wallis statistic	df	p-value
Age	50,057	29	0,009

Source Elaborated from JASP Team (2023)

different longevity values. However, these results do not go in depth into the differences between the different routes, so a comparison of their confidence intervals has been developed, which are shown in Fig. 3. For practical reasons, each route is identified by a numerical code. The list of routes and their corresponding codes is given in Annex I.

The confidence intervals show no notable differences at the general level. The discrepancy revealed by the Kruskall-Wallis test occurs between wine routes number 6 (Enoturisme Penedès) and 13 (Ribera del Duero), as well as between routes 6 and 20 (Toro). Thus, with few exceptions, there is a large overlap in the longevity of the wineries on the different routes. More specifically, most of them are over 25 years old, the most extreme case being the average of 42 years in wineries on the Calatayud route. This is a sign of the longevity of the organizations that make up most of the routes, which is a first indication of the potential historical and cultural wealth that the territories can demonstrate through wine tourism.

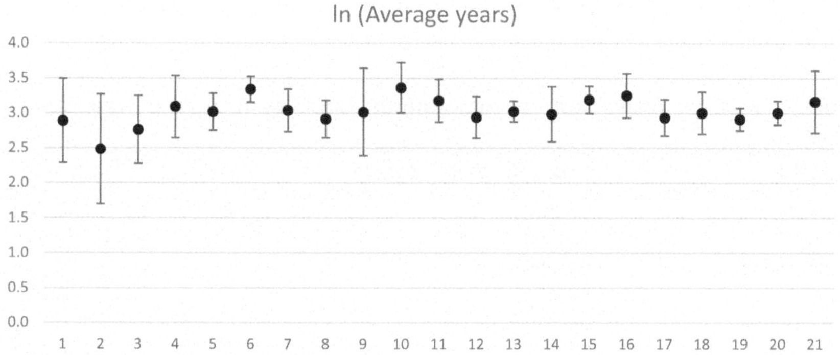

Fig. 3 95% confidence intervals on wine routes' wineries average longevity (*Source* Own elaboration)

Structure of the Wine Routes' Wineries

Similar to the previous case, an analysis of various accounting magnitudes has been carried out, both through the Kruskall-Wallis test and through the study of the overlapping confidence intervals. It should be mentioned that the various variables considered are in line with the current literature on the size-survival relationship. In essence, as postulated by a multitude of authors (Fackler et al., 2013; Fotopoulos & Louri, 2000; Kaniovski & Peneder, 2008; Talay et al., 2014; Ugur & Vivarelli, 2021), larger organizations show a higher probability of survival. The same phenomenon occurs with the age of organizations, since the more extended their activity has been over time, the less likely they are to perish (Banbury & Mitchell, 1995; Dimara et al., 2008; Esteve-Pérez & Mañez-Castillejo, 2008; Esteve-Pérez et al., 2004; Talay et al., 2014). Therefore, the study of various magnitudes that reflect this size is interesting from a strategic point of view.

Table 2 shows the results of the various Kruskall-Wallis tests that have been carried out. As can be seen, the only cases in which there are no differences between the various groups of wineries according to the route to which they are linked occur in fixed assets and number of employees (level $\alpha = 0.05$). It is therefore advisable to carry out a more in-depth analysis of these variables, since, by presenting a common pattern, it makes it possible to design strategies that are similar to a large majority of routes.

Table 2 Kruskall-Wallis test for accounting variables from the wine routes' wineries

Variable	Kruskall-Wallis statistic	Df	p-value
Current assets	56,282	29	0,002
Fixed assets	40,568		0,075
Total assets	51,099		0,007
Net sales	52,476		0,005
Employees	29,451		0,442
Net profit	63,533		< 0,001

Source Elaborated from JASP Team (2023)

According to the results, there are singularities in terms of current assets and total assets (which, in part, are composed of the former), as well as in sales and net profit. This points to the fact that the different routes offer differentiated returns, which justifies work such as that of Martínez-Falcó et al. (2023), so that it is possible to design strategies specific to each of the environments in a unique way. However, both fixed assets and the number of employees do not differ among the different routes considered. Unlike the rest of the variables, these are linked to a greater degree to strategies with high level and large temporal scope, since they are related both to investments in long-term assets and rights (fixed assets) and to productive force and knowledge (number of employees). However, the use of the Kruskall-Wallis test as a non-parametric test is more prone to the appearance of type II errors, so the above results are complemented for both variables by studying the confidence intervals. These results are given in Table 3 (fixed assets) and Table 5 (employees).

With regard to fixed assets, when computing the confidence intervals associated with the natural logarithm of the variable,[1] statistically significant differences are observed, especially in the Arribes and Bullas routes. This pattern arises because both have a lower average asset value than the rest, which is respectively €243,806 and €335,206 (Table 4). On the other hand, the rest of the routes do not show systematic discrepancies, the only notable case being Cigales route, also with a usually low value with respect to the bulk of the routes. Precisely, the vast majority of routes present a much higher average asset volume than the previous ones, which are consistently found with average resources per winery above one million euros, which denotes the fact that the vast majority of routes have wineries with an infrastructure and properties of considerable size.

In terms of the number of employees, the situation is slightly different from the previous case. With a few isolated exceptions, the only systematic divergence appears in the Arribes route, where the average number of employees per winery organization is 1.4, which denotes an eminently

[1] The use of confidence intervals requires normal distribution. Therefore, the natural logarithm transformation is used to obtain a statistical distribution with skewness and kurtosis values in the range [−1,1], which is associated with the aforementioned normal distribution.

Table 3 95% confidence overlap on fixed assets per wine route

	1	2	3	4	5	6	7	8	9	10	11	12	13	14	15	16	17	18	19	20	21
1																					
2	■																				
3	■																				
4																					
5																					
6																					
7		■																			
8			■																		
9		■	■																		
10		■	■	■																	
11		■	■																		
12		■	■																		
13		■	■	■		■															
14																					
15		■	■																		
16		■	■	■																	
17		■	■	■																	
18																					
19		■	■																		
20																					
21																					

Source Own elaboration

autonomous activity. On the contrary, the rest of the routes are above 4.7 employees, which implies, in any case, more than 330% of the value of the Arribes route. In any case, and as illustrated in Table 6, the average

Table 4 Average fixed asset value per wine route

Route	Average fixed assets (€)	Route	Average fixed assets (€)	Route	Average fixed assets (€)
1	2,996,416	8	1,464,528	15	2,384,978
2	243,806	9	1,781,020	16	3,208,133
3	335,206	10	3,657,545	17	3,801,343
4	482,668	11	1,803,449	18	1,173,759
5	1,080,196	12	1,225,462	19	1,943,373
6	995,385	13	3,247,517	20	959,918
7	1,511,129	14	1,227,941	21	1,210,784

Source Elaborated from SABI (Bureau van Dijk, 2023)

number of employees indicates that, in general terms, the bulk of the routes are composed of small wineries.

Conclusion and Discussion

In this chapter, an analysis of different variables has been carried out to characterize most of the Spanish wine routes, both in terms of geographical scope and structure. From the results obtained, which correspond to the end of the year 2021, the first conclusion is that these routes are divided into three main groups according to their geographical scope. Specifically, the greatest atomization occurs in the areas linked to the various Rioja and Ribera del Duero wine routes. On the other hand, there are average levels in the Galician, Asturian, and Leonese areas, as well as in Catalonia. The rest of the routes, which are distributed along the coast in the lower half of the Iberian Peninsula, show a much greater potential range.

With regard to the structural characteristics, the analyses carried out do not show a generic pattern, as in the previous case. However, it is worth highlighting the fixed assets and employee variables, since they have shown certain specificity problems (type II errors), although they have been detected and duly analyzed. In this case, it has been observed how the divergences are essentially centered on two routes, whose variables associated with strategic issues in the medium and long term are much lower than those of the rest of the population.

Table 5 95% confidence overlap on number of employees per wine route

	1	2	3	4	5	6	7	8	9	10	11	12	13	14	15	16	17	18	19	20	21
1																					
2	■																				
3		■																			
4		■																			
5		■																			
6		■																			
7		■																			
8		■																			
9		■																			
10		■																			
11		■	■																		
12		■																			
13		■																			
14		■	■																		
15		■																			
16		■																			
17		■									■			■							
18		■																			
19		■																			
20		■																			
21		■																			

Source Own elaboration

In view of the above results, it is concluded that, as far as structural characteristics are concerned, it is not possible to determine a series of aggregate strategic proposals, since most of the routes present intrinsic particularities that force an individualized study and design for each one

Table 6 Average number of employee value per wine route

Route[2]	Employees	Route	Employees	Route	Employees
1	9.8	9	9.2	15	9.7
2	1.4	10	12.5	16	15.9
4	4.7	11	10.0	17	11.3
5	6.8	12	8.8	18	6.9
6	8.3	13	11.2	19	4.9
7	8.6	14	8.7	20	8.1

Source Elaborated from SABI (Bureau van Dijk, 2023)

of them. However, the geographic scope does show a certain degree of systematization, so it is possible to make some suggestions about the approaches to be adopted. In this way, and following the proposals for corporate strategies, thanks to the developments of Ansoff (1957) and Abell (1980), as indicated by Guerras Martín and Navas López (2022), as well as competitive ones proposed by Porter (1990).

With regard to the group with the greatest atomization, its potential geographic scope is very limited, so that, in addition to greater difficulty in attracting customers in volume, the very presence of a greater number of routes that are configured as rivals may lead to cannibalization among them. For this reason, and in order to avoid problems of excessive competitiveness that could result in joint damage, the recommendation for this type of routes is to establish a competitive strategy of differentiation, so that the particularities of each of the territories are exploited, emphasizing the virtues of the regions and counties in which they are located. To this end, it is proposed to increase the number of visitors organically, that is, by using the resources directly associated with each of the routes.

Contrary to the previous case, routes with a medium and large geographic scope are characterized not only by a large territory of influence, but also by a smaller number of nearby routes. Thus, and combining these two realities, the strategic proposal involves the establishment of alliances with related sectors. In essence, the high geographic

[2] As opposed to the previous case, many wineries had problems with the natural logarithm transformation, which is why they have been discarded in this analysis. Consequently, and unlike the rest of the variables considered, 18 wineries are used in the analysis.

scope reduces the risk of excessive competition, which, added to a greater number of resources, makes it possible to complement the wine tourism activity itself with other services and experiences allow to offer a high-value product. An example of this could be the incorporation of museums, visits to cities or monuments, among others, so that the routes themselves are conceived as part of a larger environment that allows, in turn, to improve the understanding of its reality. In addition, offering a series of additional services that are different from the winemaking activity itself can lead to a greater number of visitors who, indirectly, can end up referring to the wine tourism activity.

In essence, as shown, there are strategic possibilities for the different groups of routes, although those with greater atomization present certain difficulties that make strategic development more stringent. However, it should be noted that this work has certain limitations. The first of these is that it is based solely on Spanish territory, so its extrapolation to other countries is subject to the similarity of the routes that exist there. In addition, this analysis is based only on data for the year 2021, so further developments are required in terms of time range to analyze the possible evolution and dynamism of the routes.

Despite the above, the authors are confident in the usefulness of these results, which, although generic, show certain lines of action from a strategic point of view that can be used for the development of wine routes. Given the aforementioned importance of these routes, due to their contribution to the economic development of areas at risk of depopulation, the proposals derived from the results are of practical use to both the private sector (for the development of strategies in winery organizations) and the public and public–private sectors (design of strategies and policies for infrastructures and tourism actions). Consequently, it can be concluded that, although there are certain limitations (as in any research work), the brevity of the results allows for a generalized analysis in certain aspects, so that a basis has been established from which to start for the design of higher-level strategies that will subsequently have to be particularized for each of the routes.

> The most important investment you can make is in yourself.
> —Warren Buffett

Annex I: Wine Routes' Identification Codes

Code	Route	Code	Route	Code	Route
1	Alicante	8	Lleida	15	Rioja Alavesa
2	Arribes	9	La Mancha	16	Rioja Alta
3	Bullas	10	Marco de Jerez	17	Rueda
4	Cigales	11	Navarra	18	Somontano
5	DO Empordà	12	Rías Baixas	19	Txakolí
6	Enoturisme Penedès	13	Ribera del Duero	20	Toro
7	Jumilla	14	Ribera del Guadiana	21	Utiel-Requena

References

Abell, D. F. (1980). *Defining the business.* Prentice Hall.

Ansoff, I. (1957). Strategies for diversification. *Harvard Business Review, 35*(5), 113–124.

Banbury, C. M., & Mitchell, W. (1995). The effect of introducing important incremental innovations on market share and business survival. *Strategic Management Journal, 16*(1), 161–182.

Bureau van Dijk. (2023). *SABI.* https://login.bvdinfo.com/R0/SabiNeo

Dimara, E., Skuras, D., Tsekouras, K., & Tzelepis, D. (2008). Productive efficiency and firm exit in the food sector. *Food Policy, 33*(2), 185–196.

Erwig, M. (2000). The graph Voronoi diagram with applications. *Networks: An International Journal, 36*(3), 156–163.

Esteve-Pérez, S., & Mañez-Castillejo, J. A. (2008). The resource-based theory of the firm and firm survival. *Small Business Economics, 30*, 231–249.

Esteve-Pérez, S., Sanchis-Llopis, A., & Sanchis-Llopis, J. A. (2004). The determinants of survival of Spanish manufacturing firms. *Review of Industrial Organization, 25*, 251–273.

Fackler, D., Schnabel, C., & Wagner, J. (2013). Establishment exits in Germany: The role of size and age. *Small Business Economics, 41*, 683–700.

Fotopoulos, G., & Louri, H. (2000). Location and survival of new entry. *Small Business Economics, 14*, 311–321.

Goss-Sampson, M. A. (2022). *Statistical analysis in JASP: A guide for students.*
Guerras Martín, L. Á, & Navas López, J. E. (2022). *La dirección estratégica de la empresa: Teoría y aplicaciones* (6th ed.). Thomson Reuters.
JASP Team. (2023). JASP (Version 0.17.3.0) [computer software].
Kaniovski, S., & Peneder, M. (2008). Determinants of firm survival: A duration analysis using the generalized gamma distribution. *Empirica, 35*, 41–58.
Marco-Lajara, B., Martínez-Falcó, J., Millán-Tudela, L. A., & Sánchez-García, E. (2023). Analysis of the structure of scientific knowledge on wine tourism: A bibliometric analysis. *Heliyon, 9*(2), e13363.
Martínez-Falcó, J., Marco-Lajara, B., Zaragoza-Sáez, P., & Sánchez-García, E. (2023). Wine tourism in Spain: The economic impact derived from visits to wineries and museums on wine routes. *Investigaciones Turísticas, 25*, 168–195. https://doi.org/10.14198/INTURI.21219
Porter, M. E. (1990). The competitive advantage of nations. *Harvard Business Review, 68*(2), 73–93.
Porter, M. E. (1998). *Clusters and Competition, on Competition, 7*, 91.
Talay, M. B., Calantone, R. J., & Voorhees, C. M. (2014). Coevolutionary dynamics of automotive competition: Product innovation, change, and marketplace survival. *Journal of Product Innovation Management, 31*(1), 61–78.
Turner, R. (2021). deldir: Delaunay Triangulation and Dirichlet (Voronoi) Tessellation [computer software].
Ugur, M., & Vivarelli, M. (2021). Innovation, firm survival and productivity: The state of the art. *Economics of Innovation and New Technology, 30*(5), 433–467.

Developing Sustainable Tourism in Tanzania: Perspective from Wine Stakeholders

Galinoma Gahele Lubawa and Evans S. Osabuohien

Introduction

The Tanzanian wine industry is a vital sector that makes a significant contribution to the country's economic, social, and cultural landscape. It is worth noting that the agriculture sector, which includes the production of grapes used to make wine, contributes significantly to Tanzania's GDP. In 2021, Tanzania's GDP grew by 4.9%, supported by the global economic recovery, with growth being driven by agriculture and services on the supply side, and final consumption and investment on the demand side (African Development Bank, 2022). Despite its importance, the industry faces numerous challenges that affect wine marketing, growth, and sustainability of wineries (Lubawa & Osabuohien, 2023).

G. G. Lubawa (✉)
Institute of Rural Development Planning, Dodoma, Tanzania
e-mail: galinoma@gmail.com

E. S. Osabuohien
Covenant University, Ota, Ogun State, Nigeria
e-mail: Pecos4eva@gmail.com; DIaDeRC.Centre@gmail.com

J. Martínez-Falcó et al. (eds.), *Wine Tourism and Sustainability*,
https://doi.org/10.1007/978-3-031-48937-2_3

To overcome these hurdles, winemakers must explore effective ways of marketing their products, with a focus on wine tourism strategies. The winery sector often considers wine tourism as a crucial component of a destination's appeal, and it can be a major motivating factor for visitors (Hall et al., 2000). Wine tourism can indeed be an opportunity for wineries to build relationships with customers and provide them with an immersive experience in the world of grapes and wine. Through wine tourism, wineries can showcase their unique qualities, including their history, culture, and the surrounding area's socioeconomic influences (World Tourism Organisation, 2021). Properly planned and managed wine tourism can provide a significant opportunity for growth in the wine value chain and benefit operators, visitors and important environmental issues (Sun & Drakeman, 2021; Tafel & Szolnoki, 2020). The presence of wine travelers in the area can ensure livelihoods for the community, support other local businesses, and help grow the tourism sector in Tanzania as a whole.

Tanzania is a recognized destination for nature-based tourism; however, wine tourism is a relatively new concept in the country. There is an opportunity for the country to launch wine tourism and achieve great success by learning from successful wine tourism destinations like Germany, Portugal, and Australia (Correia et al., 2022; Sun & Drakeman, 2021; Tafel & Szolnoki, 2020). According to the latest World Bank economic analysis, Tanzania has a unique opportunity to revamp the tourism industry to drive inclusive growth over the long term and promote sustainable and resilient tourism (World Bank, 2021). This includes promoting new forms of tourism such as wine tourism, which can contribute to the country's annual income and expand its tourism attractions. Tanzania has a thriving tourism sector, but it is important to develop a unique strategy for introducing wine tourism that considers the distinct characteristics of this market. By conducting comprehensive research, planning carefully, and implementing effectively, Tanzania could potentially create a successful wine tourism industry that would bolster economic growth and enhance the country's tourism offerings. To develop a successful wine tourism industry, Tanzania will need to assess the current state of its wine industry, determine the target market for wine tourism, and create a plan to cater to the needs of this market.

Despite the extensive research has been conducted on wine tourism in various countries around the world (Correia et al., 2022; Sun & Drakeman, 2021), however, research on wine tourism in Tanzania is limited. Existing literature in Tanzania focuses primarily on the difficulties of wine production, distribution, and factors influencing small-scale wine grape growers (Lubawa & Osabuohien, 2023). According to the government report released in May 2022 (National Bureau of Statistics, NBS, 2022), Tanzania has not yet made wine tourism one of its tourist attractions. This demonstrates the scarcity of literature on wine tourism.

Furthermore, the World Bank published a report that discusses the challenges facing the tourism sector in Tanzania and highlights the need for policy actions to ensure that the sector becomes a sustainable engine of private-sector-driven growth and social and economic development (World Bank, 2021). Similarly, tourism has been known as one of the driving forces for the socioeconomic development of destination locations (UNWTO, 2021). By strategically leveraging wine tourism and effectively removing existing barriers, Tanzania has the potential to enhance its tourism industry and foster overall growth in the sector.

Thus, this study examines the current business environment for wine tourism in Tanzania and identifies opportunities for its establishment. To achieve this objective, a qualitative research design was employed, incorporating data from stakeholders in the Tanzanian wine sector. The results of this study are essential for wine sector growth and wine tourism. Overall, the results of this study have the potential to be significant not only for Tanzania but also for other emerging countries looking to establish and develop their wine tourism industry.

The Context of Wine Tourism

Wine tourism, also referred to as *ecotourism or vinitourism* (Oltean & Gabor, 2022), is a type of gastronomy tourism where the goal of travel is to visit vineyards and wineries and taste, consume, and buy wine frequently at or close to the source of grape growing and winemaking (Hall et al., 2002). It consists of a wine-related activity that offers a dynamic, adaptable experience that integrates wine culture and heritage, to elicit feelings, sensations, attachments, and sensory impressions during the visit, allowing the wine tourist to become an advocate for that specific winery, company, or winegrowing area (Santos et al., 2019). It is a relatively new industry sector that is facing economic development challenges, including consumer satisfaction, loyalty, and behavioral intentions (Byrd et al., 2016). This tourism sector focuses on getting to know the world of wine as you travel, typically through vineyards, winery tours, and wine tastings (Hall et al., 2002). Thus, wine tourism indeed encompasses a wide range of components that contribute to a memorable experience for tourists, including infrastructure, physical spaces, landscaping, cuisine, wine sales, winery visits, hospitality, and exceptional service (Maracajã et al., 2022; Torres et al., 2021).

Wine tourism provides visitors with an opportunity to experience the unique culture, history, and traditional food pairing of wine regions, thereby offering a sensory and educational experience for travelers (Joy et al., 2021). This form of tourism is characterized by visitors' experiential engagement with food and related products and activities (Becker & Jaakkola, 2020). Food and gastronomy play significant roles in tourism destinations, as they possess the potential to attract visitors and stimulate local business development. The eight top wine areas in the world reportedly attracted more than 40 million foreign wine tourists in 2016 (Mintel Group Ltd., 2017).

A study by Santos et al. (2021) also found four dimensions of tourism wine storytelling, wine tasting thrill, wine engagement, and wine landscape can measure the overall behavior of wine tourists. Wine tourism plays a critical role in promoting local and rural development by preserving resources and contributing to the sustainability and durability of the economic system, providing a platform for wine producers and local communities to promote their products and cultural heritage, supporting their economic growth, and contributing to the development of local communities and economies (Kastenholz et al., 2023). Thus, wine tourism can promote local economic development through new business opportunities, wealth creation, and job creation, ultimately improving the quality of life of local people (Hojman et al., 2012).

Wine tourism has become an increasingly popular business strategy with over forty million annual visitors to vineyards. Wine tourism has benefits for the neighboring areas that go beyond economic ones. However, this economic benefit may come at the cost of increasing carbon emissions (Sun & Drakeman, 2021). The wine industry in the Americas, including the United States, Argentina, and Chile, has experienced significant growth and increased its presence in various socioeconomic sectors through market-adapted strategies. The development of wine tourism is one of these strategies, which has led to a positive economic impact on these countries (García-Rodea et al., 2022). According to a report by United Nations World Tourism Organisation (UNWTO, 2021), the wine industry is a significant contributor to the global tourism sector, and wine tourism is a key segment of this industry. The report highlights the Wine Routes of the World, which takes the reader through the wine routes of the top ten wine producer destinations, including South Africa, Chile, Argentina, and the United States. Thus, it is a business that has the potential to impact the economy of the concerned nation. It is a significant tourist attraction with plenty of generate economic, social, cultural, and environmental benefits that can boost any country's economy (Robinson, 2023).

The Theoretical Perspective

Stakeholder Theory and Wine Tourism Sustainability

The development of sustainable wine tourism is an important goal for many communities, as it can create opportunities for economic diversification and long-term sustainability while preserving the natural and cultural resources of the regions involved. One such strategy for achieving this goal is through interaction with different stakeholders to create opportunities for economic development in emerging wine regions (UNWTO, 2021). This approach is based on the concept of sustainable tourism. It aims to achieve a balance between the economic, social, and environmental features of visiting the attractions' progress and ensure sustainability (Ridoff et al., 2022). Therefore, the stakeholder theory and sustainability are important aspects of wine tourism.

The stakeholder theory is a management framework that emphasizes the importance of considering the interests of all stakeholders in a business, not just shareholders or investors. According to Byrd, (2007), the stakeholder theory proposes that organizations should consider the interests of all stakeholders, including the host community, visitors, and future generations when making decisions. In wine tourism, stakeholders include winery owners, managers, visitors, and the host community. Sustainability in wine tourism involves practices that aim to preserve natural resources and ensure the durability of the economic system (Alonso et al., 2022).

To find the best way to establish wine tourism in Tanzania as a new entry in the tourism sector, the study has used stakeholder theory to involve potential stakeholders from the grape value chain (Fig. 1) to get their ideas (Kulwijila et al., 2018). The stakeholder theory is suggested to be a useful framework for involving all important stakeholders in the development of wine tourism in Tanzania, including government officials, local communities, and wine producers (Festa et al., 2020). By involving stakeholders in wine tourism development, Tanzania can create a framework that fosters cooperation, exchange, and cooperation between various stakeholders. Furthermore, involving stakeholders in the planning and development of wine tourism can help to ensure that the sector benefits local communities and contributes to sustainable economic growth (World Bank, 2021; UNWTO, 2021).

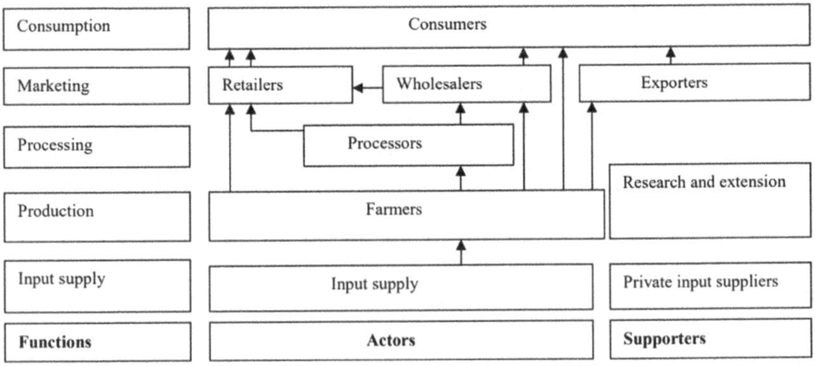

Fig. 1 Grape value chain map (*Source* Kulwijila et al. [2018])

According to a study by Sun and Drakeman (2021), wine tourism can provide economic benefits to local communities, but also has environmental costs, such as increasing carbon emissions in Australia. It is becoming increasingly popular as a business strategy, with 40 million tourists visiting wineries each year. The study reveals that tourism contributes to over 33% of the wine industry's carbon footprint, and a trade-off analysis of emissions for revenue.

Festa et al. (2020) noted that wine routes are crucial for governance and management, but facing opportunities and challenges in Italy. The study highlighted the need for better collaboration and communication between stakeholders to enhance the development of wine tourism in small municipalities. The findings also suggested that the development of wine tourism in Italy could be improved by increasing the quality of local products and services and by investing in sustainable tourism practices. Furthermore, the study identified the need for local governments to support wine tourism development through targeted policies and incentives. Overall, the study provided valuable insights into the opportunities and challenges facing the wine tourism industry in Italy and offered recommendations for stakeholders to enhance its development.

Poitras and Donald (2006) emphasize the importance of community involvement in decision-making processes to build trust and promote the long-term sustainability of the industry. Overall, the study underscores the need for a holistic approach to wine tourism development that balances economic benefits with social and environmental considerations. Community participation is indeed a critical element of sustainable wine tourism, and involving the host community as one of the stakeholders is essential to ensure the long-term sustainability of the industry (Poitras & Getz, 2006). Community participation is one of the four key dimensions of sustainability in wine tourism. The other three dimensions are environmental sustainability, economic sustainability, and cultural sustainability (Poitras & Donald, 2006). It is important to note that involving the community in decision-making processes can help build trust and promote the long-term sustainability of the industry. According to one study, the main drivers for sustainable wine tourism are to build trust and reputation, increase visitor satisfaction, and make the world a better place (Ridoff et al., 2022). In addition, the study also identified the main barriers to sustainable wine tourism, which include a lack of financial resources, a lack of human resources, and a lack of infrastructure.

In sum, the development of sustainable wine tourism requires a multi-stakeholder approach that considers environmental, economic, and socio-cultural factors. By promoting emotional connections between wine tourists, regions, and producers and following sustainable tourism principles, it is possible to create opportunities for economic development while preserving the natural and cultural resources of the communities involved.

Design of the Study

A qualitative research using in-depth interviews with Tanzania wine stakeholders in Dodoma City was engaged in the study. Qualitative research is an inductive approach that is commonly used in studies with less available information and aims to gain a deeper understanding of reality (Lubawa & Osabuohien, 2023). This approach involves examining how individuals perceive social and economic realities, and data is usually collected through semi-structured interviews (Tenny et al., 2022). For this particular study, the researchers collected data through in-depth interviews with wine sector stakeholders until data saturation was reached. In-depth interviews involve conducting detailed interviews with a small number of participants, and unlike other forms of qualitative research, researchers using an in-depth interviewing approach invest a significant amount of time with each participant using a conversational format (Rutledge et al., 2020). Since literature is scarce on wine tourism, particularly on tourism visitors ' experience in Tanzania, a qualitative approach is considered an appropriate starting point for this study (Lubawa & Osabuohien, 2023).

Data Collection and Mode of Analysis

The study engaged winery owners, wine producers, visitors, and wine sellers from the city of Dodoma as participants. Purposive sampling was used to select the 20 respondents who participated in the study, including *Wineries owners, grape growers, and wine sellers (retailers and wholesalers)* who are knowledgeable about the wine sector in Tanzania. Their experience in the value chain, and the exposure they got in this sector, will enable them to give them contribute ideas on how to start and maintain the wine tourism sector. A sample size of 20 was chosen to establish data saturation and provide adequate responses to the main research questions (Baker & Edwards, 2016). The study was conducted from February to April 2023. Respondents were visited in their work environment. The study was completely voluntary, and participants were able to stop participating at any time. Questions were asked about the development and sustainability of wine tourism products. Each interview lasted thirty to forty-five minutes.

A systematic coding process was carried out to analyze the data, based on three main themes: Establishment, Products, and Sustainability. Braun and Clarke (2020) define thematic analysis as a technique for assessing qualitative data that involves searching through the data to find recurrent themes, interpreting them, and summarizing the findings. The first stage of coding the information collected uses open coding. Selective coding takes the data one step further by combining it into major themes, and identifying the key findings from the dataset, and axis coding takes the data one step further by grouping like-answers to further consolidate the information (Lubawa & Osabuohien, 2023; Braun & Clarke, 2006).

Results and Discussion

Stakeholders' Profile

This section presents information on the general characteristics of owners in terms of experience, gender, and experience in the grape value chain as a farmer, wine processor, or seller. The interviews were conducted with the proprietors of 8 wineries, 6 grape farmers, and 6 wine sellers (retailers and wholesalers) (Table 1).

Table 1 Profile of participants

Respondent	Position of interviewee	Gender	Experience in the grape value chain
1	Winery owner	Male	15 years
2	Grape grower	Male	18 years
3	Wine seller	Male	10 years
4	Winery owner	Female	14 years
5	Wine seller	Male	8 years
6	Wine seller	Female	12 years
7	Winery owner	Male	14 years
8	Wine seller	Female	9 years
9	Grape grower	Male	20 years
10	Winery owner	Male	7 years
11	Grape grower	Male	11 years
12	Winery owner	Female	10 years
13	Grape grower	Male	17 years
14	Wine seller	Male	7 years
15	Winery owner	Male	12 years
16	Grape grower	Female	7 years
17	Winery owner	Male	13 years
18	Winery owner	Female	5 years
19	Wine seller	Male	6 years
20	Grape grower	Female	8 years

Source The Authors'

Stakeholder Perspectives on Developing Wine Tourism and its Sustainability

For this study, we used the Braun & Clarke (2006) thematic analysis approach to produce a thematic map with quotation extracts, code sets, and concluding themes (see Appendix I). As a result of this study, four final themes have been generated, which are *Landscape, Vineyard, Cultural heritage, and Wine Festival* (Table 2).

Landscape

According to the stakeholders, Dodoma's landscape has the potential to draw tourists. This landscape is essential for wine tourism as it provides a unique and attractive place for tourists to experience the winemaking

Table 2 Extracted quotation of the answers proposed to the question "If Tanzania were to start wine tourism, what areas or products would make it attractive?"

Extracted quotation	Coding	Theme
If Tanzania were to start wine tourism, what would make it sustainable and attractive?		
Grapes are grown on naturally fertile soils. The nature of the countryside can be said to be attractive, so food with a natural taste is born	• Wine tourism is driven by Dodoma Region's unique landscape, people, history, and traditions • Bantu-speaking Gogo, also known as Wagogo, are people who live in central Tanzania • Grapes grown in Dodoma are unique and attractive • Wine from naturally fertile land • In exploring the complex wine-producing rural territory, its specific landscapes, people, history, and traditions, are recognized as the main drivers for wine tourism • Tourists can be attracted to natural food, and a pleasant environment to relax in the vineyards • Dodoma can be called a "wine city." • The presence of wine concerts is recommended • The aim is to make Dodoma wine known to people inside and outside Tanzania • Wine represents culture	1. Landscape 2. Vineyard 3. Cultural heritage 4. Wine Festival

(continued)

Table 2 (continued)

Extracted quotation	Coding	Theme
The climate in Dodoma is generally savannah-like, with a long dry season from late April to early December and a brief wet season in the other months. High amounts of evaporation and soil erosion occur due to persistent dry winds, low humidity, and other factors		
Visiting vineyards offers a variety of activities, both indoors and outdoors, with good weather and fresh air		
The gogo tribal people have a unique traditional way of life, with their gastronomy (in gogo translated as "*ndigwa*"), dances, and mud houses (in gogo translated as "itembe" or matembe)		
Dodoma-produced wine has a sweet taste. So, after visiting national parks perhaps in *Mikumi, Serengeti National Park, Tarangire National Park, and Lake Manyara National Park,* tourists can come to Dodoma for some relaxation, taste wine, and eat people's traditional food. This is also tourism		
Dodoma is one of the few cities that has been well planned through urban design, so the presence of vineyards and the investment in wine factories allow the city to be promoted as a "wine city."		

(continued)

Table 2 (continued)

Extracted quotation	Coding	Theme
The introduction of the Dodoma Wine Festival...... during the wine festival, together with traditional dances and other types of entertainment, is anticipated to boost Dodoma's economy		
You know wine is like a natural drink. Every country started the production of wine in natural ways. Thus, we can keep our original ways to attract other nations to come and experience our culture		

process. This involves visiting vineyards and wineries to taste different wines, learn about farming processes, and enjoy the surroundings. In particular, one stakeholder said;

> ... the vineyards are like natural vegetation, green enough for tourism. They are interesting to watch and breathe fresh and natural air. A tourist who has finished visiting a national park can come to the farm to rest and relax.

Globally, the vineyard landscape provides a compelling backdrop for guests and enriches the entire wine tourism experience. These landscapes are characterized by vineyards, rolling hills, and beautiful landscapes. Mitchell et al., (2012) indicated that wine tourism could be explored from a non-tourist viewpoint, considering the wine landscape. In Australia's Margaret River region, pastoralism has influenced wine tourism, while France has experienced a cultural environment with historical and social value due to its role as the location of wine production. Land tenure regimes have also impacted champagne wine tourism.

Since wine tourism and the landscape are closely related, this study suggests that Tanzania should create a distinctive environment for tourists to enjoy while learning about grape farming there. Grape growers, winemakers, and the community as a whole could benefit from this. This may be possible as a result of Tanzania receiving a high number of foreign visitors, with an estimated 1.6 million visitors spending US$2.5 billion and contributing to more than 17% of the GDP, around 30% of foreign earnings, and 10% of formal employment (Anderson et al., 2021). Vineyard sceneries are produced as a result of wine production, serving as both natural and man-made landscapes that draw travelers to these picturesque areas. Due to the immersive experience, they offer guests, these vineyard landscapes are essential to the wine tourism industry (Winkler & Kimberly, 2016).

Vineyard

Wine stakeholders claim that vineyards (Fig. 2) may contribute to wine tourism in Tanzania. Visitors can explore the terroir of the area and learn about the grape-growing process, as well as tasting several wines. Vineyards can provide tours, tastings, food and wine pairings, and unique experiences. These experiences can help visitors gain a greater understanding of the craft of winemaking and the distinctive flavors and aromas of various wines. Weather is one of the main obstacles to viticulture, but Tanzania's fine weather also allows the vineyards to be included in wine tours (Säfwenberg, 2019). One of the stakeholders claimed that;

> ... without a vineyard, wine tourism cannot be fully experienced. It is a place where visitors can find out about the grape-growing process, see the many grape varieties that are used to make various wines, and experience terroir, or the specific combination of soil, temperature, and topography that gives each wine its unique flavour and aroma.

Another respondent said;

> ... a weekend getaway might include relaxing in the vineyard's setting. Therefore, if made appropriately, it might be a draw for travellers looking to unwind while enjoying the relaxing atmosphere of vineyards.

Fig. 2 Vineyards outside the town of Dodoma, Tanzania (*Source* The Citizen [2021, December 3])

Vineyards play a crucial role in the wine tourism industry through activities like tours and tastings. Visitors can learn about the grape-growing process and sample different wines while enjoying the beautiful scenery of the vineyards. Additionally, some vineyards offer food and wine pairing experiences, allowing visitors to learn how to pair local cuisine with the vineyard's wines (Crespi-Vallbona et al., 2020). Visitors can participate in the grape harvest and winemaking season, gaining a deeper appreciation for the art of winemaking. (Haller et al., 2021). Thus, visitors can learn about the winemaking process, taste different wines, and appreciate the beauty of the vineyards in Tanzania.

Cultural Heritage

Another finding of this study indicates that stakeholders relate wine tourism to cultural heritage. Stakeholders have suggested that cultural events like traditional dances (e.g., *masumbi, muheme*), traditional food (called "*ndigwa*" in the gogo language), and traditional clothing that are found in the Dodoma region can make wine tourism (which is referred to as "*nhembelo ye cing'wegwa co mnhogoni*" in gogo tribe), more interesting, unique, and attractive. Dances, clothing, and some types of food continue to be part of the heritage of this tribe and can be enjoyed at various festivals. Stakeholders advise combining wine tourism with other cultural pursuits to draw in more international visitors, particularly those who are interested in discovering Tanzanian culture. One of the respondents said;

> ... Tourists can combine wine tourism with traditional dancing. Visitors to the vineyards can enjoy dances such as the marimba, zeze and muheme (which he named after the dance kigogo) while they are there. For a small entrance fee, dance group members can earn money.

Literature suggested that wine tourism combines elements of cultural and gastronomic tourism (Sun & Drakeman, 2021). Wine is central to wine tourism, and it involves the discovery of wine, terroir, and the people of the region in which they are located (Gastronomy and Wine Tourism I UNWTO, 2020) (Fig. 3).

Fig. 3 Wagogo playing a traditional dance (*Source* Dodoma FM [2023])

Wine Festival

To encourage local wine production and bring tourists interested in Tanzanian wines, it was suggested that the Dodoma Wine Festival should be organized yearly. Wine tourism has the potential to boost local economies and create jobs. For instance, wine tourism can generate revenue from wine tastings, tours, and other wine-related activities. Additionally, wine tourism can also create employment opportunities and support local businesses, such as restaurants, hotels, and souvenir shops (Sun & Drakeman, 2021). One of the stakeholders explained that;

> ... If the wine festival is organised at least once a year, it can draw sponsors and gain recognition both inside and outside of Tanzania. By providing local entertainment, as a nation, we may hope to promote the event and encourage participation. For attendees to have a fantastic day, we can offer networking opportunities, live music, and company growth. A range of foods, such as barbecue and local cheese, can be served at the event.

Studies suggest that wine festivals have the potential to generate significant economic benefits for the regions in which they are held. The economic benefits of wine festivals can take many forms, including increased tourism, job creation, and revenue for local businesses. A study published by Kruger et al. (2019), examined the economic impact of the Finger Lakes Wine Festival in New York. The study found that the festival generated $15.3 million in economic activity and supported 180 jobs in the region. The study also found that attendees of the festival had high levels of satisfaction with their experience and were likely to return in the future. A study conducted in South Africa by Joubert, Saayman, and Saayman (2013) found that wine festivals played an important role in promoting wine sales in the Western Cape region. Another study conducted by Saayman et al. (2012) found that wine festivals in South Africa significantly impacted regional tourism and the promotion of regional wines.

Martínez-Falcó et al. (2023) utilized structural equation modeling to investigate the influence of Corporate Social Responsibility (CSR) on

Green Innovation Performance (GIP) in the Spanish wine sector. While the study did not focus on the specific economic consequences of the winery and museum visits on vineyard pathways, it contributes significant insights to the overall understanding of sustainability in the Spanish wine sector.

Tanzania has the opportunity to host an annual sustainable wine festival in close collaboration with key wine industry players. The festival could serve as a powerful platform to boost the growth and development of the wine industry while boosting the country's overall tourism sector. Valuable lessons can be learned from South Africa in the Western Cape's Robertson Wine Valley. In the region, the two influential wine festivals, Robertson Wacky Wine Weekend and Robertson Wine on the River, stimulated the expansion of the wine tourism industry and helped raise awareness of the region's wines among a wider audience (Joubert et al., 2013; Saayman et al., 2012).

Conclusion

The study findings identified landscape, vineyards, cultural heritage, and wine festivals as potential wine tourism attractions in Tanzania. These products showcase the country's natural beauty, cultural richness, and winemaking expertise, and can serve as drawcards for both local and international tourists. Tanzania receives 1.6 million tourists annually, which generates 2.5 billion dollars in revenue, 17% of the country's GDP, 30% of foreign income, and 10% of formal employment (Anderson et al., 2021). Therefore, Tanzania would experience numerous economic impacts if the tourist industry is expanded by launching wine tourism.

Therefore, it is recommended that policymakers and tourism stakeholders in Tanzania focus on leveraging these products to establish a thriving wine tourism industry. This can be achieved through targeted marketing campaigns, investment in infrastructure and facilities, and the implementation of sustainable practices that preserve and enhance the natural and cultural resources of the country. By doing so, Tanzania can maximize the economic benefits of wine tourism while ensuring the long-term sustainability of the sector.

References

African Development Bank. (2022). African Economic Outlook 2022. Supporting Climate Resilience and a Just Energy Transition in Africa. ISBN 978–0–9635254–2–0; African Economic Outlook 2022. African Development Bank - Building today, a better Africa tomorrow (afdb.org).

Anderson, W., Mossberg, L., & Andersson, T. (Ed.). (2021). *Sustainable tourism development in Tanzania*. Cambridge Scholars Publishing.

Alonso, D. A., Bressan, A., Kok, K. S., & O'Brien, S. (2022). Filling up the sustainability glass: Wineries' initiatives towards sustainable wine tourism. *Tourism Recreation Research, 47*(5–6), 512–526. https://doi.org/10.1080/02508281.2021.1885801

Asili ya ngoma za wagogo – Dodoma FM. (2023, April 14). https://radiotadio.co.tz/dodomafm/2023/04/14/asili-ya-ngoma-za-wagogo/

Baker, S. E., & Edwards, S. (2016). How many qualitative interviews are enough? Expert voices and early career reflections on sampling and cases in qualitative research. Retrieved June 15, 2016, from http://eprints.ncrm.ac.uk/2273/4/how_many_interviews.pdf

Becker, L., & Jaakkola, E. (2020). Customer experience: Fundamental premises and implications for research. *Journal of the Academy of Marketing Science, 48*(4), 630–648. https://doi.org/10.1007/s11747-019-00718-x

Braun, V., & Clarke, V. (2006). Using thematic analysis in psychology. *Qualitative Research in Psychology, 3*(2), 77–101. https://doi.org/10.1191/1478088706qp063oa

Byrd, E. T., Canziani, B., (Jerrie) Hsieh, Y.-C., Debbage, K., Sonmez, S. (2016). Wine tourism: Motivating visitors through the core and supplementary services. *Tourism Management, 52*, 19–29.https://doi.org/10.1016/j.tourman.2015.06.009

Byrd, E. T. (2007). Stakeholders in sustainable tourism development and their roles: Applying stakeholder theory to sustainable tourism development. *Tourism Review, 62*(2), 6–13. https://doi.org/10.1108/16605370780000309

Braun, V., & Clarke, V. (2020). One size fits all? What counts as quality practice in (reflexive) thematic analysis? *Qualitative Research in Psychology*. https://doi.org/10.1080/14780887.2020.1769238

Carlsen, J. (2004). A review of global wine tourism research. *Journal of Wine Research, 15*(1), 5–13. https://doi.org/10.1080/0957126042000300281

Correia, A. I., Melo, C., & Carvalho, C. (2022). Wine tourism experiences: An exploratory approach to innovation potential in the Douro Valley, Portugal. In J. V. D. Carvalho, P. Liberato, A. Peña (Eds.), *Advances in tourism, technology and systems. Smart innovation, systems and technologies* (Vol. 284). Springer. https://doi.org/10.1007/978-981-16-9701-2_20

Crespi-Vallbona, M., & Mascarilla-MirÃ, O. (2020). Wine lovers: Their interests in tourist experiences. *International Journal of Culture, Tourism and Hospitality Research, 14*(2), 239–258. https://doi.org/10.1108/ijcthr-05-2019-0095

Festa, G., Shams, S. M. R., Metallo, G., & Cuomo, M. T. (2020). Opportunities and challenges in the contribution of wine routes to wine tourism in Italy–A stakeholders' perspective of development. *Tourism Management Perspectives, 33*, 100585. https://doi.org/10.1016/j.tmp.2019.100585

García-Rodea, L. F., Thomé-Ortiz, H., Espinoza-Ortega, A., & de Alcântara Bittencourt-César, P. (2022). Viniculture and tourism in the new world of wine: A literature review from the American continent. *Wine Economics and Policy, 11*, 127–140. https://doi.org/10.36253/wep-10897

Gastronomy and Wine Tourism I UNWTO. (2020). https://www.unwto.org/gastronomy-wine tourism

Haller, C., Hess-Misslin, I., & Mereaux, J.-P. (2021). Aesthetics and conviviality as key factors in a successful wine tourism experience. *International Journal of Wine Business Research, 33*(2), 176–196. https://doi.org/10.1108/IJWBR-12-2019-0063

Hall, M., Longo, A. M., Mitchell, R., & Johnson, G. (2002). Wine tourism in. *Wine tourism around the world: Development, management and markets*, 150.

Hall, C. M., Sharples, L., Cambourne, B., & Macionis, N. (2000). *Wine Tourism Around the World: Development, management and markets*. Elsevier Butterworth-Heinemann.

Hojman, D. E., & Hunter-Jones, P. (2012). Wine tourism: Chilean wine regions and route. *Journal of Business Research, 65*(1), 13–21. https://doi.org/10.1016/j.jbusres.2011.07.009

Joubert, J., Saayman, M., & Saayman, A. (2013). Wine tourism and sales at wine festivals: A case study in the Western Cape Province, South Africa. *Tourism Management, 36*, 248–254.

Joy, A., Yoon, S., Grohmann, B., & LaTour, K. (2021). How winery tourism experience builds brand image and brand loyalty. *Wine Business Journal*. https://doi.org/10.26813/001c.30210

Kastenholz, E., Paço, A., & Nave, A. (2023). Wine tourism in rural areas – hopes and fears amongst residents. *Worldwide Hospitality and Tourism Themes, 15*(1), 29–40. https://doi.org/10.1108/WHATT-08-2022-0095

Kulwijila, M., Makindara, J., & Laswai, H. S. (2018). Grape value chain mapping in Dodoma Region, Tanzania. *Journal of Economics and Sustainable Development, 9*, 171–182.

Kruger, M., & Viljoen, A. (2019). Terroir wine festival visitors: Uncorking the origin of behavioural intentions. *Current Issues in Tourism, 1–21*. https://doi.org/10.1080/13683500.2019.1667310

Lubawa, G. G., & Osabuohien, E. S. (2023). Wine Industry in Tanzania and its Sustainability: Qualitative Insights from Family-Owned Wine Processing Enterprises Owners. In B. Marco-Lajara, A. Gilinsky, J. Martínez-Falcó, & E. Sánchez-García (Eds.), *Handbook of research on sustainability challenges in the Wine Industry* (pp. 194–235). IGI Global. https://doi.org/10.4018/978-1-6684-6942-2.ch011

Martínez-Falcó, J., Marco-Lajara, B., Zaragoza-Sáez, P. D., & Millán-Tudela, L. A. (2023). Analyzing the effect of Corporate Social Responsibility on Green Innovation Performance in the Spanish wine industry: A structural equation modelling analysis. *Agribusiness*.

Maracajá, K., Schramm, V. B., Schramm, F., Valduga, V., & F., B. (2022). A multicriteria model for evaluation of Brazilian wineries from a tourism destination perspective. *International Journal of Wine Business Research, 34*(1). https://doi.org/10.1108/IJWBR-12-2020-0057

Mintel Group Ltd. (2017). Wine tourism. https://www.aaaa.org/wp-content/uploads/2017/04/Global-Mintel-Wine-tourism-2017-03.pdf

Mitchell, R., Charters, S., & Albrecht, N. J. (2012). *Cultural Systems and the Wine Tourism Product, 39*(1), 311–335. https://doi.org/10.1016/j.annals.2011.05.002

National Bureau of Statistics (NBS). (2022). *Press release: Tourist arrivals for the period of January to May 2022*. https://www.nbs.go.tz/nbs/Takwimu/; https://www.nbs.go.tz/nbs/takwimu/Tourism/PRESS_RELEASE_TOURISTS_ARRIVALS_JAN_TO_MAY%202022.pdf.

Oltean, F. D., & Gabor, M. R. (2022). Wine tourism—A sustainable management tool for rural development and vineyards: Cross-cultural analysis of the consumer profile from Romania and Moldova. *Agriculture, 12*(10), 1614. https://doi.org/10.3390/agriculture12101614

Poitras, L., & Donald, G. (2006). Sustainable wine tourism: The host community perspective. *Journal of Sustainable Tourism, 14*(5), 425–448. https://doi.org/10.2167/jost587.0

Ridoff, N., Nilsson, C., Szolnoki, G., Tafel, M., & Stelter, A.-C. (n.d.). (2022). *Sustainable Wine Tourism. A global survey.* https://www.winetourism.com/files/2022/01/Sustainable_wine_tourism_Report_GeisenheimUni_WineTourismcom.pdf

Robinson, D. M. (2023). From sustainability to regeneration: The role of wine tourism in building back better wine regions. In B. Marco-Lajara, A. Gilinsky, J. Martínez-Falcó, & E. Sánchez-García (Eds.), *Handbook of research on sustainability challenges in the wine industry* (pp. 32–52). IGI Global. https://doi.org/10.4018/978-1-6684-6942-2.ch003

Rutledge, P. B., & Hogg, J. L. C. (2020). In-depth interviews. *The International Encyclopedia of Media Psychology, 1–7.* https://doi.org/10.1002/9781119011071.iemp0019

Säfwenberg, M. (2019). Svenskt vin: En vinvärld som växer. Idus förlag, Sverige. Tourism in Skåne AB (2011, March 22). Skåne - Europas nya vindistrikt. Press release from Tourism in Skåne AB. http://www.mynewsdesk.com/se/tourisminskane/pressreleases/skaane-europas-nyavindistrikt-600969

Saayman, M., Saayman, A., & Joubert, J. (2012). The impact of wine festivals on the economy of the Cape Winelands District Municipality. *Journal of Travel & Tourism Marketing, 29*(1), 44–55.

Santos, V., Ramos, P., Sousa, B., & Valeri, M. (2021). Towards a framework for the global wine tourism system. *Journal of Organizational Change Management, 35*(2), 348–360. https://doi.org/10.1108/JOCM-11-2020-0362

Santos, V. R., Ramos, P., Almeida, N., & Santos-Pavón, E. (2019). Wine and wine tourism experience: A theoretical and conceptual review. *Worldwide Hospitality and Tourism Themes, 11*(6), 718–730. https://doi.org/10.1108/whatt-09-2019-0053

Sun, Y., & Drakeman, D. (2021). The double-edged sword of wine tourism: The economic and environmental impacts of wine tourism in Australia. *Journal of Sustainable Tourism.* https://doi.org/10.1080/09669582.2021.1903018

Tafel, M., & Szolnoki, G. (2020). Estimating the economic impact of tourism in German wine regions. *International Journal of Tourism Research*, jtr.2380. https://doi.org/10.1002/jtr.2380

Torres, J. P., Barrera, J. I., Kunc, M., & Charters, S. (2021). The dynamics of wine tourism adoption in Chile. *Journal of Business Research, 127*, 474–485. https://doi.org/10.1016/j.jbusres.2020.06.043

Tenny, S., Brannan, J. M., & Brannan, G. D. (2022). *Qualitative study.* StatPearls Publishing.

The Citizen. (2021, December 3). *Dodoma grows into the wine capital of East Africa.* https://www.Thecitizen.co.tz. Retrieved May 2, 2023, from https://www.theeastafrican.co.ke/tea/magazine/dodoma-grows-into-the-wine-capital-of-east-africa-3638420

United Nations World Tourism Organization (UNWTO). (2021, September 15). Gastronomy and wine tourism. https://www.Unwto.org/Gastronomy-Wine-Tourism. Retrieved April 29, 2023, from https://www.unwto.org/gastronomy-wine-tourism

World Bank. (2021). Transforming tourism: Toward a sustainable, resilient, and inclusive sector. https://documents.worldbank.org/en/publication/documentsreports/documentdetail/794611627497650414/transforming-tourism-toward-a-sustainable-resilient-and-inclusive-sector

Winkler, K. J., & Nicholas, K. A. (2016). More than wine: Cultural ecosystem services in vineyard landscapes in England and California. *Ecological Economics, 124*, 86–98. https://doi.org/10.1016/j.ecolecon.2016.01.013

Spain's Wine Tourism: Evaluating the Economic Contributions of Wineries and Museums

Javier Martínez-Falcó, Bartolomé Marco-Lajara, Eduardo Sánchez-García, and Luis A. Millán-Tudela

Introduction

Recent trends indicate a shift in tourism patterns towards briefer, more frequent excursions. Travelers are not only seeking traditional beach destinations but also desire unique experiential journeys (Martínez-Falcó et al., 2023a). This shift in preferences has led both public and private

J. Martínez-Falcó (✉) · B. Marco-Lajara · E. Sánchez-García ·
L. A. Millán-Tudela
University of Alicante, Alicante, Spain
e-mail: Javier.falco@ua.es

B. Marco-Lajara
e-mail: bartolome.marco@ua.es; bartolome.marco@gcloud.ua.es

E. Sánchez-García
e-mail: eduardo.sanchez@ua.es

L. A. Millán-Tudela
e-mail: luisantonio.millan@ua.es

J. Martínez-Falcó et al. (eds.), *Wine Tourism and Sustainability*,
https://doi.org/10.1007/978-3-031-48937-2_4

sectors to restructure their offerings, introducing thematic tourism products that emphasize the cultural, social, and culinary heritage of regions (Sánchez-Cañizares & López-Guzmán, 2012). Notably, gastronomy, especially the wine domain, has garnered significant interest among tourists.

Wine tourism epitomizes a harmonious intersection, both enriching tourists' experiences and bolstering the socioeconomic and environmental progress of viticultural regions (Marco-Lajara et al., 2023). It provides an immersive sensory journey where tourists, irrespective of their oenological expertise, can savor the multifaceted aspects of wine, from its aroma and taste to the ambiance of wineries and the ceremonial uncorking (Stewart et al., 2008). Beyond mere leisure, wine tourism invigorates both rural and urban economies while championing sustainable development. Moreover, it embeds itself in the socio-cultural tapestry, shaping the "wine landscape" and preserving local heritage (Brunori & Rossi, 2001).

Wine tourism is ascendant in Spain and traditionally robust in New World nations like the United States, Australia, Chile, and South Africa. A cluster approach, manifested through wine routes, facilitates cooperation and synergy among industry players. These routes are pivotal for wine tourism's efficacy. In Spain, scholarly investigations underscore the significance of wine routes in enhancing regional competitiveness, amplifying wine production, elevating residents' quality of life, and championing environmental stewardship (Portela & Domínguez, 2020). To the best of our understanding, the economic ramifications of these routes within the Iberian nation have not been explored in scholarly literature. To bridge this academic void, an analysis of tourism demand in relation to the economic impact of these routes is undertaken in this book chapter. Consequently, the posed research question is: What economic impact is attributed to each of these routes in the Spanish wine context?

The chapter is organized in the following manner. After this brief introduction, section "Tourism and Wine: The Intrinsic Interplay" delves into the interplay between wine and tourism, emphasizing wine routes as the linchpin of wine tourism, particularly in the Spanish context. In section "Methodology", the adopted methodology is elucidated.

Section "Results" unveils the findings, while section "Conclusions" offers conclusions, acknowledges limitations, and suggests prospective research avenues.

Tourism and Wine: The Intrinsic Interplay

Initial investigations in wine tourism can be traced back to the 1990s, centering on wine tourism's impact on rural regions (Hall, 1996). Notably, seminal contributions in this domain include *Wine Tourism Around the World* (Hall et al., 2000) and *Explore Wine Tourism* (Getz, 2000). Their groundbreaking nature in the field is acknowledged. Additionally, studies by Charters and Ali-Knight (2002), Carlsen (2004), Getz and Brown (2006), and Mitchell and Hall (2006) are recognized for their significance.

The conceptualization of wine tourism lacks uniformity, having been approached from diverse viewpoints. Hall et al. (2000) define it as the experience linked to vineyard visits, wineries, and wine presentations, with tasting as a central component. Conversely, it is perceived by Getz and Brown (2006) as a strategy for regional development and a platform for wineries to market directly to consumers. It extends beyond mere wine tasting, encompassing socio-cultural and environmental dimensions that cater to tourists seeking unique experiences. An exhaustive review of various definitions and interpretations of the term was undertaken by Clemente-Ricolfe et al. (2012).

Regarding wine tourists, two methods have traditionally been employed in scholarly literature for analysis and segmentation: (a) demographic-based categorizations (factors such as origin, age, education, and familial background); and (b) profiling based on psychographic attributes (values, attitudes, and lifestyles). Various studies incorporate both demographic and psychographic determinants, frequently characterizing wine tourists as individuals aged between 30 and 50, of moderate to high income, residing near or within the wine region, being habitual wine enthusiasts, possessing intermediate to advanced wine knowledge, and exhibiting diverse motivations for winery visits (Martínez-Falcó et al., 2023b). For this study, those visiting the wineries and wine museums affiliated with the Wine Routes of Spain are regarded as wine tourists.

Over recent decades, wine tourism's role as an auxiliary driver for employment and economic prosperity has been recognized. Consequently, a rich tapestry of academic inquiry has evolved, featuring studies from nations such as Australia, Canada, and Chile (e.g., Sigala, 2019; Torres et al., 2021). In Spain's context, early investigations concentrated on regions like La Rioja and Marco de Jerez, initially led by foreign academics. Subsequent enhancements were contributed by Spanish scholars, with notable studies spanning regions from Catalan to Ribera del Duero, produced by researchers such as Armesto-López and Gómez-Martín (2004), Medina and Tresserras (2008), Alonso and O'Neill (2009), among others.

The attraction to this typology of tourism is attributed primarily to three elements: (1) the enhancement of tourist influx within specific locales, (2) the creation of an image synonymous with a premium travel destination, and (3) its utility as an instrument for regional growth. Presently, myriad research avenues exist in this domain. In this sense, Gómez et al. (2019) have pinpointed seven distinct research trajectories concerning wine tourism.

1. Territorial Evolution: This research trajectory probes the association between wine tourism and regional economic progression. Specifically, considerations are drawn on wine tourism's residual and potential contributions to regional sustainability and development, comparative analyses of wine tourism destinations, and the significance of wine tourism in shaping regional and national brand identities.
2. Oenological Pathways: Research in this domain primarily scrutinizes the inception and dispersion of wine routes across nations. Notable examinations include those from Chile, Greece, Italy, Portugal, South Africa, and Spain. Recent advancements in this research trajectory delve into intricate and novel networks, clusters, and alliances within wine areas.
3. Wine Tourist Dynamics: Within this realm, efforts to comprehend the demography and psyche of wine tourists are paramount. Analyses

focused solely on demographic attributes and those merging demographic with psychographic facets have been conducted. Furthermore, post-visit and purchasing behaviors have been investigated to foster brand allegiance.

4. Enrichment through Winery and Tasting Encounters: Within this segment, the emphasis has been placed on the exploration of the wine tasting and winery experiences. Notably, post-visit sales, representing a pivotal distribution conduit for wineries, are closely tied to the well-being of wine tourists. Investigations encompass the aspirations and requirements of wine tourists during winery visits, alongside the quality of service dispensed.
5. Conceptual Frameworks in Wine Tourism: Within this research domain, intricate conceptual paradigms concerning wine tourism have been devised. The causal relationships between constructs pertinent to wine tourism behaviors, such as involvement and territorial affinity, juxtaposition of destination and self-image, and the interplay between tourist experience and emotional ties have been elucidated.
6. Exploration of Wine-centric Events and Festivals: Research in this domain delves into the effects of wine events and festivals on viticultural regions. Furthermore, the drivers behind visitors' attendance and the scrutiny of festival pathways are examined.
7. Marketing Strategies in Wine Tourism: Investigations have delved into consumers' regional perception dynamics, scrutiny of content disseminated on winery websites, and explorations of winery sales patterns.

In the research parameters set by Gómez et al. (2019), emphasis is placed on tourist routes, particularly examining Spain's wine route institutions' progression and the economic implications of such tourism in Spain over time. Wine routes, characterized by prescribed itineraries within specific regions, highlighting both wineries and historical landmarks, have been recognized as potent instruments for regional economic and social advancement (Hall et al., 2000). Subsequent sections delve into the literature, discussing wine route conceptualizations and their significance in both New and Old World contexts, with an accentuated focus on Spain.

Wine Tourism Routes

As alluded to earlier, a primary research trajectory in wine tourism focuses on the genesis and expansion of wine routes. In Italian legislation, a comprehensive definition of these routes has been formulated, characterizing them as marked pathways, highlighted by distinct signage, emphasizing the natural, cultural, and environmental attributes, vineyards, and accessible wineries, whether independent or cooperative (Pastor, 2006, p. 198). These routes effectively function as promotional tools for viticulture regions and their products. Consequently, wine routes foster collaborations among various tourism types, notably rural (accentuating countryside lodging), gastronomic (experiencing local cuisines), and cultural tourism (offering scenic vistas and access to local cultural landmarks).

The inception of a wine route is underpinned by dual objectives. Primarily, the image of the viticultural region and its constituent wineries is enhanced. Concurrently, such routes facilitate augmented sales for wineries, becoming pivotal for wine SMEs facing challenges accessing broad distribution channels (Fuentes-Fernández et al., 2022). To actualize a wine route, collaborative networks among service providers are imperative. Even though the wine tourist perceives a unified experience, multiple stakeholders (encompassing restaurants, hotels, and wineries) contribute. Thus, laying down legal foundations for inter-stakeholder agreements becomes crucial, necessitating an integrated body of public and private entities to champion, nurture, and oversee this tourism niche.

Central components of the tourist experience on a wine route encompass winery visits, vineyard sampling, museum admissions, local gastronomy indulgence, landscape appreciation, wine procurement, and accommodation provisions (Martínez-Falcó et al., 2023c). A direct correlation has been identified: the richer the array of tourist amenities in a region (inclusive of wineries, museums, and natural parks), the higher the incidence of tourist overnight stays. A recurrent critique identified in wine tourism literature is the limited supplementary offerings within wine routes, resulting in diminished overnight stays in wine zones (García et al., 2010).

Within the Old World, wine tourism has been organized around tourist routes, encompassing both natural vistas and tangible attractions like wineries and museums. While a significant portion of scholarly contributions on wine tourism originates from Australia and New Zealand, these Oceanic nations haven't embraced this tourism model as extensively as their European counterparts (López-Guzmán et al., 2013). This disparity is attributed to the more entrenched tradition of wine route design in Old World nations (García et al., 2010), including Italy's La Strada del Vino, France's Routes des Vin, Portugal's Rota do Vinho, and Spain's Ruta del Vino. Further elaboration on these European routes will be provided in subsequent sections.

As articulated by Charters (2009), wine tourism in the Old and New World is approached distinctly. In the Old World, a reactive stance is adopted; given the centuries-old wine production legacy of these countries, the emphasis of wine tourism often gravitates towards its rich history, cultural heritage, and culinary context. Conversely, the New World embodies a proactive approach, championing innovative and dynamic wine tourism products and the exploration of fresh markets. Here, wine tourism is seen as an integral experience and a pivotal wine distribution avenue. In contrast, the Old World's offerings predominantly center on the wine itself, viewing it as a supplementary venture to the primary production of wine (Martínez-Falcó et al., 2023d).

The Wine Routes of Spain

The genesis of wine tourism in Spain remains ambiguous. Initially, these routes were mere maps pinpointing winery locations. A significant evolution can be traced back to the 1970s, when developed tourist routes emerged in La Rioja, subsequently spreading to other viticultural regions of Spain. Spain's inaugural documented wine route is attributed to Cambados (Clemente-Ricolfe et al., 2012). Nevertheless, the requisite design and methodology for shaping the tourism product were only formalized in 1994 with the inception of the Spanish Association of Wine Cities (ACEVIN, for its acronym in Spanish). Prior to this, wine routes had been crafted predominantly in rural zones by entities with limited expertise in the field. The foundational aims of ACEVIN encompassed the organization of wine routes, orchestration across varying routes, and the cultivation of a high-caliber tourism product to amplify Spain's cultural heritage via its vinous traditions.

In 2000, an initiative titled the "Integral Plan for the Quality of Spanish Tourism" was launched by the Spanish government, targeting the enhancement of tourism product quality. Among the envisaged tourism endeavors was the "Wine Routes of Spain," conceived to bolster tourism diversification. As per the plan, a wine route was characterized as "an amalgamation of resources and tourist services, both existing and potential, from a viticultural region, anchored in authenticity and experiential intent, devised to shape a product resonating with the destination's identity, support collaborative marketing efforts across the region, and ensure visitor satisfaction, thereby fostering the comprehensive socioeconomic growth of the region" (State Secretariat for Tourism, 2000). This initiative not only bolstered the domestic and global reputation of prominent Spanish wine conglomerates but also augmented wine sales via tourist purchases within the wineries. By 2009, the Spanish government's dedication to wine route advancement materialized in the gastronomic endeavor "Savoring Spain," designed to catalyze the formation of new culinary-centric tourist locales and elevate the promotion and marketing of these destinations.

For a Wine Route in Spain to receive certification, it must navigate three distinct phases: initiation, maturation, and stabilization (López-Guzmán et al., 2013). Initially, during the initiation stage, collaborating public and private entities form a governing body to oversee the route. Adaptations are made to meet the quality criteria set forth by the General Secretariat of Tourism, concluding with the attainment of official Wine Route certification. The maturation phase ensues post-certification and spans a year, during which the oversight body ensures compliance with the General Secretariat's stipulations. A favorable evaluation allows the route to progress, whereas an unfavorable one results in certification loss, accompanied by a year's hiatus before reapplication eligibility is restored. The stabilization phase involves recurrent quality assessments. If discrepancies emerge, a six-month grace period is afforded for rectifications.

Successful amendments preserve the certification, while failures lead to its revocation and a subsequent one-year reapplication prohibition.

Currently, the tourism offering "Wine Routes of Spain" comprises 35 distinct routes, spanning 13 out of Spain's 17 autonomous regions. The procedural methodology for the inception and eventual certification of these wine routes is delineated in the Wine Routes of Spain Tourism Product Manual, authored by the referenced institutions. This methodology unfolds in three stages.

1. Initiation: During this stage, requisite standards are expected to be met by the participating public and private entities. The establishment of a governing body for the route is necessitated, a steward for this entity is designated, and facility adaptations of prospective members are aligned with the quality criteria stipulated by the overseeing institutions.
2. Advancement: Lasting two years, this stage commences upon securing the Spanish Wine Route certification. Throughout this interval, adherence to the guidelines set forth in the "Wine Routes of Spain Product Manual" is imperative for route members. Following this phase, a favorable assessment ensures retention of the certification, progressing to the subsequent stage. In contrast, failing to meet the standards results in certification revocation, precluding reapplication for a year as a consequence. Such a mechanism ensures sustained commitment to upholding and elevating requisite quality benchmarks.
3. Consolidation: In order to maintain the level of demand of the wine routes, the standards set out in the aforementioned manual are checked every two years. If, after the assessment, a negative result is obtained, the destination loses the distinction of Wine Route of Spain, having to reapply for the distinction when it meets the minimum requirements established.

Since the inception of the certification, six routes have been stripped of the "Wine Route of Spain" accolade due to non-compliance with essential quality benchmarks. These encompass the Tacaronte-Acertejo route (2009), Tenerife and Condado de Huelva (2012), Ribeiro (2013), and the Empordà and Txacolí routes (2018). Tacaronte-Acertejo and Tenerife's decertifications were rooted in the mismanagement of the defunct Tenerife Wine Route brand, an offshoot of the original Tacoronte-Acentejo (Delgado, 2016). These routes failed to validate externally the stringent quality standards mandated by the Wine Routes of Spain and neglected the execution of the approved action plans for Tenerife (Delgado, 2016). The routes of Condado de Huelva, Ribeiro, Empordà, and Txacolí faced similar quality standard lapses. Notably, efforts are underway in Condado de Huelva for re-certification (Lanza, 2022), and the Txacolí route has been recertified in 2023 (Peciña, 2021). The broader community's demand for the reinstatement of these routes underscores the prestigious image and caliber synonymous with the "Wine Routes of Spain" distinction. Furthermore, since their 2008 debut, the number of members in Spanish wine routes has experienced a commendable upward trajectory.

Methodology

To address the research inquiries, an exploratory quantitative methodology is employed, given the novelty of the subject and its absence in prior academic literature. A non-experimental design is also embraced, meaning that the independent variables remain unaltered, and the observed conditions reflect the natural setting (Garcia-Fernandez & Cordero-Borjas, 2010).

For this research, secondary information served as the primary data source. In our data-rich era, the skill to harness and interpret such data is increasingly crucial (Andrews et al., 2012). As highlighted by Hakim (1982), leveraging secondary sources fosters enriched insights. The Spanish Wine Tourism Observatory (OTVE, for its acronym in Spanish) was utilized to examine the evolution of route members (supply), tourist numbers per route (demand), and average expenditures. Emphasis was placed on analyzing route partners, particularly wineries and wine museums, given the specific data availability from OTVE. Although OTVE reports have been available since 2008, the essential data for impact calculations was only provided starting in 2017.

Hence, while earlier reports lacked detailed data, post-2017 publications offer comprehensive insights conducive to this research's objectives. The methodologies for determining the wine routes' economic impact are delineated as follows:

1. Economic Impact from Winery Visits (EIWV) is calculated by multiplying the number of winery visitors on the route with the average visitation cost and then adding the product of the number of winery visitors on the route and the average in-shop expenditure per visitor.
2. Economic Impact from Wine Museum Visits (EIWMV) is determined by multiplying the number of museum visitors on the route with the average entry cost, further supplemented by the product of the number of museum visitors on the route and the average in-museum retail expenditure per visitor.
3. The Total Economic Impact is the summation of EIWV and EIWMV.

Results

The Spanish wine routes, with their rich history and cultural significance, have long been a subject of economic interest. Recent data spanning from 2017 to 2022 offers a nuanced understanding of the economic vitality of these routes, revealing both convergent and divergent trends (see Table 1).

The Alicante route exemplifies the resilience and adaptability inherent in many of these routes. After commencing with an economic contribution of 1,127,214 euros in 2017, it witnessed a crescendo, culminating at 1,988,025 euros by 2022. However, the journey was not without its troughs, as evidenced by a dip in 2020, where the impact receded to 689,075 euros. In contrast, routes such as Arribes have maintained a more modest economic footprint. This route's contributions have remained relatively subdued over the years, with figures oscillating around the 50,000 to 70,000 euros mark. The overarching

Table 1 Economic impact of the Spanish wine routes

Ruta	2017	2018	2019	2020	2021	2022
Alicante	1,127,214	1,407,202	1,632,203	689,075	1,504,297	1,988,025
Arlanza	186,463	480,150	456,781	267,169	60,716	1,165,837
Arribes	–	–	52,222	23,180	21,744	71,117
Bierzo	421,378	429,538	433,514	68,382	204,098	235,979
Bullas	240,028	276,345	292,797	117,266	366,143	283,100
Calatayud	3,069,200	3,490,856	3,630,114	919,877	587,899	3,471,055
Campo de Cariñena	243,392	312,220	330,583	126,591	195,983	262,252
Cigales	382,739	410,610	439,489	32,583	99,513	223,017
DO Empordà	4,947,108	–	–	–	–	–
Enoturisme Penedès	13,137,776	14,099,442	12,256,858	3,689,358	7,380,603	10,456,652
Garnacha—Campo de Borja	628,470	709,050	697,199	279,941	131,685	795,719
Jumilla	850,582	1,512,439	1,721,432	380,817	1,169,613	1,582,352
Lleida	1,650,514	2,799,983	1,475,654	459,407	969,637	1,066,922
La Mancha	–	–	473,954	197,939	1,266,355	561,438
La Manchuela	–	–	813,350	147,537	317,858	149,216
Madrid	–	–	–	403,273	147,806	1,674,729
Marco de Jerez	17,704,135	16,183,501	20,986,822	2,909,626	1,615,438	11,710,907
Montilla-Moriles	364,359	349,721	481,425	71,810	8,005,066	691,139
Méntrida-Toledo	–	–	–	–	–	50,922
Navarra	288,604	580,502	703,199	232,223	374,844	635,442
Rías Baixas	2,228,187	2,625,707	2,686,072	868,575	2,315,242	3,798,738
Ribera del Duero	7,112,841	13,451,407	11,820,742	3,330,709	2,508,350	8,995,898
Ribera del Guadiana	1,441,713	802,272	844,831	123,787	10,080,254	483,639
Rioja Alavesa	6,171,087	6,842,669	6,898,516	2,397,556	246,142	8,435,390
Rioja Alta	7,099,723	10,218,654	11,718,972	2,483,723	1,424,215	11,034,239
Rioja Oriental	–	121,291	156,762	44,983	10,369,429	61,660

Ruta	2017	2018	2019	2020	2021	2022
Ronda	1,108,400	543,165	994,615	124,378	63,526	640,855
Rueda	991,504	1,333,797	1,388,020	233,649	350,757	1,211,807
Sierra de Francia	19,057	29,269	46,440	17,786	1,333,767	129,665
Somontano	2,173,474	2,333,105	4,923,202	1,344,171	99,867	2,077,340
Txakolí	66,266	–	–	–	2,361,389	127,294
Toro	–	–	1,287,059	456,207	2,208,675.98	2,058,040
Utiel-Requena	1,436,572	1,730,689	1,845,966	687,591	2,302,383	2,023,557
Valdepeñas	–	–	174,740	56,175	1,506,175	140,102
Yecla	179,177	159,341	135,572	44,294	69,915	161,051
Zamora	–	–	–	–	–	12,429
Total	75,269,963	83,073,585	91,799,104	23,229,637	61,659,386	78,467,523

Source Prepared by the authors based on the OTVE

narrative, though, is one of growth and potential. From a collective economic contribution of 75,269,963 euros in 2017, the wine routes have managed to nearly maintain this momentum, registering 78,467,523 euros by 2022. The dip in 2020, with a total of 23,229,637 euros, serves as a testament to the external challenges the industry faced, yet also underscores its resilience. Furthermore, Fig. 1 shows that this economic impact derives to a greater extent from visits to wineries than from visits to wine museums.

Regarding the wine routes with the greatest impact, first, the Marco de Jerez route emerges as a paragon of economic resilience and robustness. Its peak in 2019, with a staggering contribution nearing 21 million euros, is a testament to its unparalleled significance. Even in the face of adversities, as witnessed in 2020, this route demonstrated an unwavering commitment to economic vitality. Second, Enoturisme Penedès stands as another beacon of consistent economic impact. Its sustained performance over the years, culminating in a substantial contribution in 2022, accentuates its pivotal role in the Spanish wine tourism sector. Third, the Ribera del Duero route, with its impressive figures, especially in the year 2018, underscores its central role in driving economic activity. Its contributions have been instrumental in bolstering the overall economic health of the sector. Fourth, the Rioja Alta route, with its remarkable

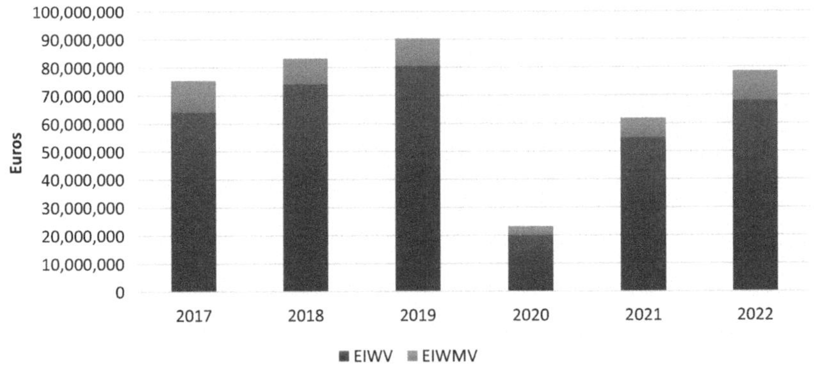

Fig. 1 Composition of the economic impact of the wine routes (2017–2022) (*Source* Prepared by the authors based on the OTVE)

contributions in years like 2019 and 2022, further cements its position as a major economic pillar.

As regards the routes with lower impact, first, the Arribes route consistently demonstrates a restrained economic impact. Its contributions over the years have hovered around modest figures, such as 52,222 euros in 2017 and 71,117 euros in 2020. Second, the Méntrida route, with an economic impact of just 50,922 euros in one of the years, further exemplifies the routes that have a lesser economic influence in the broader landscape of Spanish wine tourism. Third, the Sierra de Francia route also finds itself on the more modest end of the economic spectrum. With figures like 19,057 euros in one year and 129,665 euros in another, its contributions, while valuable, are comparatively modest. Fourth, the Zamora route, with its contribution of just 12,429 euros in one of the years, further underscores the routes that have a more limited economic impact.

Conclusions

The findings of this study hold significant relevance for scholars, industry stakeholders, and practitioners within the wine domain, enhancing the depth of understanding surrounding Spain's wine tourism sector. This investigation sheds light on the intricacies of both the provision and consumption patterns within the Spanish wine routes, elucidating the trajectory of affiliated entities and the fiscal ramifications of wine tourism on Spanish vineyards and related museums. By delving into this niche, the study addresses a previously uncharted territory in scholarly discourse, offering fresh insights into the economic dimensions of wine-centric activities in Spain.

Spanish vineyards have recognized the pivotal role of wine tourism in elevating the stature of their wines both domestically and internationally. Furthermore, wine tourism seamlessly intertwines with other tourism facets, encompassing culinary, cultural, and wellness experiences (notably vinotherapy). This confluence fosters a rich tapestry of synergistic opportunities, amplifying the overall appeal and impact.

The "Wine Routes of Spain" initiative presents a multifaceted array of advantages that cater to diverse stakeholders, from consumers to entrepreneurs, and even to the overarching regulatory framework. For consumers, this initiative offers an immersive experience, allowing them to delve into the milieu where the wine originates and benefit from an enriched and expertly curated service. Entrepreneurs, on the other hand, gain from the establishment of a robust and unique collective brand identity. This collaborative venture fosters mutual support, taps into an evolving market segment, mitigates seasonal demand fluctuations, and introduces advanced service quality protocols, educational opportunities, centralized procurement, and cohesive promotional strategies. Lastly, these routes facilitate a structured approach to enotourism. Local municipalities, regulatory bodies, and the very entrepreneurs steering these wine routes collaboratively sculpt and brand the tourism offering, ensuring a harmonized and impactful experience for all.

Emerging from the analysis is the undeniable influence of the COVID-19 pandemic on the viticulture tourism sector. Both the supply chain and consumer demand metrics reveal a precipitous downturn. The pandemic's onslaught has left an indelible mark on the Spanish viticulture landscape. While domestic wine consumption witnessed an uptick, avenues such as the Horeca channel, direct vineyard sales, and enotourism bore the brunt of the pandemic's repercussions (Marco-Lajara et al., 2021). The mobility restrictions instituted during the pandemic's peak in 2020 led to a contraction in institutional participation in the wine routes by 2.58%. Concurrently, visitor engagement with the various Spanish wine routes plummeted by 73.53%, and the fiscal contribution of the sector receded by 74.7% relative to the preceding year.

The study elucidates distinct variations among the Spanish wine routes concerning economic contributions. Notably, routes like Marco de Jerez and Enoturisme Penedès emerge as linchpins of economic vitality, reflecting their central role in shaping the fiscal discourse of Spain's wine tourism. Their unwavering prominence over the years is indicative of their profound economic significance, intertwined with cultural and touristic allure. In contrast, routes such as Arribes, Méntrida, and Sierra de Francia, though holding intrinsic value, have established themselves as nuanced contributors. Their steadfast contributions over time beckon a deeper inquiry into the diverse determinants shaping their economic landscape. The broader narrative of the wine routes, characterized by a predominant growth trajectory interspersed with sporadic deviations, portrays an industry marked by change and boundless potential. The timeless charm of Spanish enotourism, augmented by its nimbleness, heralds its stature as a promising economic domain. Distilling these observations, the narrative of the Spanish wine routes emerges as a tapestry of vigor, tenacity, and promise, underscoring the symbiotic relationship between culture, tourism, and fiscal dynamics in Spain's core.

Several determinants elucidate the heterogeneous contribution of each Spanish route. Key among these are the enotourism offerings of the respective vineyards, their geographical positioning relative to major tourist hubs, the developmental trajectory of the destinations, the brand stature of the locales and their wines, and the accessibility afforded by transportation linkages to these wine routes.

A salient managerial insight emerging from the study suggests that vineyard proprietors would benefit substantially from aligning with the wine routes encompassing their establishments. Engaging in enotourism under the aegis of a recognized wine route can confer myriad benefits, including amplification of on-site wine sales, enhancement of brand prestige, and the opportunity for activity diversification. Concurrently, the Wine Routes of Spain's tourism observatory ought to consistently disseminate information regarding its wine tourism offerings, fostering initiatives that elevate the visibility of the collective brands under its purview. Furthermore, the findings of this study can serve as a strategic compass for decision-makers contemplating investments, be it through acquisitions or direct capital infusion, particularly in vineyards situated along the more economically influential routes.

While this study offers valuable insights, it is imperative to acknowledge its inherent constraints. The research primarily focuses on the economic repercussions stemming from visits to vineyards and museums. Consequently, a more holistic impact of enotourism in the encompassed destinations might be more expansive when considering facets such as lodging, culinary establishments, and activity enterprises. Moreover, the reliance on data from the OTVE could potentially skew the derived figures. Historically, there's been an inclination towards underreporting wine tourist numbers, as not every visitor to vineyards and museums is duly accounted for. To address these constraints, future research endeavors could aim to broaden the economic impact analysis to encompass all stakeholders within the wine routes. Additionally, a direct engagement with vineyard proprietors could shed light on the pandemic's influence on their wine tourism ventures and gather insights on the efficacy of the measures introduced by the Spanish administration to bolster the viticulture industry.

References

Alonso, A., & O'Neill, M. (2009). Wine tourism in Spain: The case of three wine regions. *Tourism: An International Interdisciplinary Journal, 57*(4), 405–420.

Andrews, L., Higgins, A., Andrews, M., & Lalor, J. (2012). Classic grounded theory to analyse secondary data: Reality and reflections. *The Grounded Theory Review, 11*(1), 12–26.

Armesto-López, X., & Gómez-Martín, B. (2004). Productos agroalimentarios de calidad, turismo y desarrollo local: El caso del Priorat. *Cuadernos Geográficos, 34*, 83–94.

Brunori, G., & Rossi, A. (2001). Synergy and coherence through collective action: Some insights from wine routes in Tuscany. *Sociology Ruralis, 40*, 1–19.

Carlsen, P. (2004). A review of global wine tourism research. *Journal of Wine Research, 15*(1), 5–13.

Charters, S. (2009). New world and Mediterranean wine tourism: A comparative analysis. *Tourism: An International Interdisciplinary Journal, 57*(4), 369–379.

Charters, S., & Ali-Knight, J. (2002). Who is the wine tourist? *Tourism Management, 23*(3), 311–319.

Clemente-Ricolfe, J., Escribá-Pérez, C., Rodriguez-Barrio, J., & Buitrago-Vera, J. (2012). The potential wine tourist market: The case of Valencia (Spain). *Journal of Wine Research, 23*(2), 185–202.

Delgado, R. (2016). *De espaldas al enoturismo de calidad*. https://www.eldiario.es/canariasahora/premium-en-abierto/enoturismo-calidad-vino-canarias-ruta_1_4227882.html

Fuentes-Fernández, R., Martínez-Falcó, J., Sánchez-García, E., & Marco-Lajara, B. (2022). Does ecological agriculture moderate the relationship between wine tourism and economic performance? A structural equation analysis applied to the ribera del duero wine context. *Agriculture, 12*(12), 2143.

Garcia-Fernandez, F., & Cordero-Borjas, A. (2010). Proceso de gestion del conocimiento en Carabobo (Venezuela) y Tamaulipas (Mexico). *Pensamiento & Gestión, 28*, 132–154.

García, J., Guzmán, T., Ruiz, S., & García, M. (2010). Turismo del vino en el marco de Jerez. Un análisis desde la perspectiva de la oferta. *Cuadernos de Turismo, 26*, 217–234.

Getz, D. (2000). *Explore wine tourism, management, development and destinations*. Nueva. Cognizant Communication Corporation.

Getz, D., & Brown, G. (2006). Critical success factors for wine tourism regions: A demand analysis. *Tourism Management, 27*(1), 146–158.

Gómez, M., Pratt, M., & Molina, A. (2019). Wine tourism research: A systematic review of 20 vintages from 1995 to 2014. *Current Issues in Tourism, 22*(18), 2211–2249.

Hakim, C. (1982). *Secondary analysis in social research: A guide to data sources and methods with examples*. Allen and Unwin/Unwin Hyman.

Hall, C. (1996). *Wine tourism in New Zealand*. Proceedings of tourism down under II: A research conference (pp. 109–119). Donedin: Universidad de Otago.

Hall, C., Sharples, L., Cambourne, B., & Macionis, N. (2000). *Wine tourism around the world: Development, management and markets*. Editorial Elsevier.

Lanza. (2022). *Rosa Melchor y el alcalde de La Palma del Condado se reúnen para poner en marcha una nueva ruta del vino*. https://www.lanzadigital.com/provincia/melchor-el-alcalde-de-la-palma/

López-Guzmán, T., García, J., & Rodríguez, Á. (2013). Revisión de la literatura científica sobre enoturismo en España. *Cuadernos de Turismo, 32*, 171–188.

Marco-Lajara, B., Seva-Larrosa, P., Martínez-Falcó, J., & Sánchez-García, E. (2021). How Has COVID-19 Affected The Spanish Wine Industry? An Exploratory Analysis. *Natural Volatiles & Essential Oils Journal*, 2722–2731.

Marco-Lajara, B., Úbeda-García, M., Zaragoza-Sáez, P., Poveda-Pareja, E., & Martínez Falcó, J. (2023). Enoturismo y sostenibilidad: Estudio de casos en la Ruta del Vino de Alicante (España). *PASOS Revista de Turismo y Patrimonio Cultural, 21*(2), 307–320.

Martínez Falcó, J., Marco-Lajara, B., Zaragoza-Sáez, P., & Sánchez-García, E. (2023a). *Vino, Turismo y COVID-19: El impacto de la COVID-19 en las Rutas del Vino de España*.

Martínez-Falcó, J., Marco-Lajara, B., Zaragoza Sáez, P. D. C., & Sánchez-García, E. (2023b). Wine tourism in Spain: The economic impact derived from visits to wineries and museums on wine routes. *Investigaciones Turísticas, 25*, 168–195.

Martínez-Falcó, J., Marco-Lajara, B., Zaragoza-Sáez, P., & Millan-Tudela, L. (2023c). Do circular economy practices moderate the wine tourism–green performance relationship? A structural analysis applied to the Spanish wine industry. *British Food Journal* (ahead-of-print No. ahead-of-print).

Martínez-Falcó, J., Marco-Lajara, B., Zaragoza-Sáez, P., & Millan-Tudela, L. A. (2023d). Wine tourism as a catalyst for green innovation: evidence from the Spanish wine industry. *British Food Journal* (ahead-of-print No. ahead-of-print).

Medina, F. (2015). Tourism and culture in names food and wine origin: The case of the region Tokaj (Hungary). *International Journal of Scientific Management and Tourism, 1*(3), 167–177.

Medina, F., & Tresserras, J. (2008). Wine tourism and wine routes in Catalonia. Case analyses: DO Penedes, DO Priorat and DO Montsant. *PASOS: Revista de Turismo y Patrimonio Cultural, 6*(3), 493–509.

Mitchell, R., & Hall, C. M. (2006). Wine tourism research: The state of play. *Tourism Review International, 9*(4), 307–332.

Pastor, E. (2006). *El turismo del vino, otra experiencia de ocio.* Bilbao: Editorial Deusto.

Peciña, M. (2021). *Ayala, a punto de recuperar el certificado de la Ruta del Txakoli.* https://www.elcorreo.com/alava/araba/ayala/ayala-punto-recuperar-20211103163553-nt.html

Portela, J., & Domínguez, M. (2020). Las rutas del vino como motores de dinamización socio-territorial: El caso de Castilla y León. *Boletín de la Asociación de Geógrafos Españoles, 84*, 1–36.

Sánchez-Cañizares, S., & López-Guzmán, T. (2012). Gastronomy as a tourism resource: Profile of the culinary tourist. *Current Issues in Tourism, 15*(3), 229–245.

State Secretariat for Tourism. (2000). *Propuesta para la asistencia técnica para el desarrollo del estudio de definición del producto "Rutas del Vino de España."* Servicio de Publicaciones del Ministerio de Industria, Comercio y Turismo.

Sigala, M. (2019). Building a wine tourism destination through coopetition: The business model of ultimate winery experiences Australia. In *Wine Tourism Destination Management and Marketing* (pp. 99–112). Palgrave Macmillan.

Stewart, J., Bramble, L., & Ziraldo, D. (2008). Key challenges in wine and culinary tourism with practical recommendations. *International Journal of Contemporary Hospitality Management, 20*(3), 303–312.

Torres, J., Barrera, J., Kunc, M., & Charters, S. (2021). The dynamics of wine tourism adoption in Chile. *Journal of Business Research, 127*, 474–485.

Environmental Impact of Wine Tourism

Sustainable Wine Tourism: Best Practices

Dhruv Kishore Bole

Introduction

Wine tourism, like any other industry, has its drawbacks. It has contributed significantly to GHG (Greenhouse Gas) emissions, resulting in climate change and changes in weather patterns, decimating vineyards, and making grape vine production more difficult (Sun & Drakeman, 2021). Poor weather patterns caused by industry GHG emissions have also resulted in early grape ripening, which has jeopardized vintners' stylistic nature of the wine, resulting in lower consumer satisfaction and, ultimately, loss of commercial potential for winemakers (Chrobak & Zimmer, 2022). Loss of economic opportunities has, in turn, jeopardized the livelihoods of industry members, ultimately affecting the wine tourism business severely. As a result of such challenges, questions have been raised about the viability of wine regions around the world, as

D. K. Bole (✉)
Faculty-Food and Beverage Service, State Institute of Hotel Management, Siddhpur, Gujarat, India
e-mail: dhruv.bole@sihm.ac.in

J. Martínez-Falcó et al. (eds.), *Wine Tourism and Sustainability*,
https://doi.org/10.1007/978-3-031-48937-2_5

well as the nation's future as a wine tourist destination (Jones & Webb, 2010). The wine tourism business is fragmented and diverse, with a web of interconnected participants ranging from small- and large-scale winemakers to grape growers, wine tour operators, local community, and wine regulatory bodies. It is up to these wine tourism sector participants to ensure the survival of wine regions and the country as a wine tourist destination. These participants may make wine tourism sustainable by implementing sustainable business practices and actively working with one another (Filopoulos & Frittella, Designing sustainable and responsible wine tourism experiences, 2019). Wineries and wine tour operators, two major participants in the wine tourism industry, may be a driving force in making the wine tourism sector more sustainable through their sustainable business models (Szolnoki et al., 2022).

Sustainable wine tourism is built on three pillars: social, economic, and environmental, and it contributes to the restoration of the region's environmental, economic, and social balance (Dang, 2017; Szolnoki et al., 2022) (Fig. 1). Sustainable wine tourism requires winemakers and wine tour operators to use sustainable and responsible business practices that help reduce GHG emissions, meet local community needs and aspirations, encourage local tourism, preserve tourists' health and well-being, and conserve natural resources as well as the region's cultural legacy (The Porto Protocol, 2021).

It is imperative to note that sustainable wine tourism is becoming increasingly popular among tourists worldwide. Tourists are increasingly preferring sustainable travel and would like to experience sustainable wine destinations (Amarando et al., 2019; Montella, 2017). This is due to the fact that tourists, particularly millennials and Gen Z, are concerned about the environmental, climate, and socioeconomic repercussions of the things they buy, including wine (Pomarici & Vecchio, 2014). Sustainably minded wine connoisseurs, a tourist group in particular, are curious about vineyards' sustainability practices, as well as how environmentally, economically, and socially sustainable their favorite brand is (Amarando et al., 2019).

Sustainable wine tourism extends beyond traditional wine tasting sessions and winery and vineyard excursions. A well-planned sustainable tourism activity provides tourists with an immersive experience with

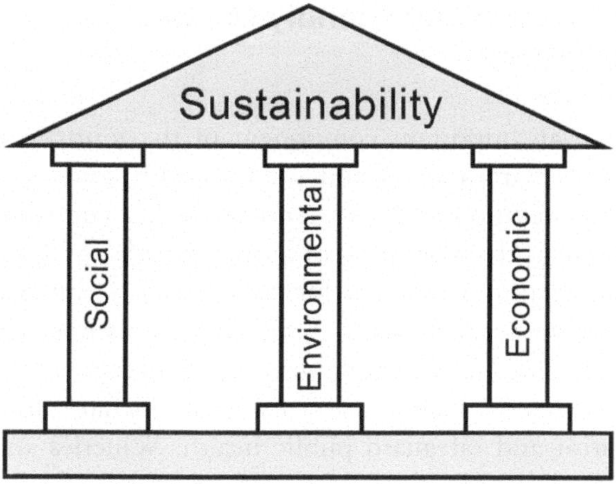

Fig. 1 Pillars of sustainability (Dang, 2017)

environment, exposure to local custom and culture, and real-life interaction with the surrounding community. Such substantial initiatives will boost the tourist experience of the destination, consequently affecting the influx of tourists in the region. It is crucial to emphasize that any region's wine tourism can be considered sustainable if it can both maintain and attract new wine tourists (Karagiannis & Metaxas, 2020). A healthy stream of visitors to the wine regions would eventually generate revenue opportunities for winemakers, wine tour operators, and neighboring communities, paving the path for regional and national development. Sustainable wine tourism could thus spur the growth of wine regions all over the world. It would also aid in shifting the public perception of the region from one of mass tourism to one of sustainability. The following part addresses the sustainable practices that wineries and wine tour operators can use to integrate sustainability into their core business activities, so contributing to the sustainability of wine tourism.

Using Environmentally Friendly Modes of Transportation

Transportation, an important component of the tourism industry, has major negative environmental and public health consequences. It has contributed to pollution and CO_2 emissions that contribute to global warming, as well as the worsening of human health problems (Stefanica, 2017). Transportation accounted for 37% of CO_2 emissions in 2021, accounting for more than half of all CO_2 emissions (International Energy Agency, 2022). Adopting sustainable modes of transport is a simple approach for wineries and wine tour operators to lower their carbon footprint and safeguard public health. Wineries and wine tour operators should consider providing environmentally friendly forms of transportation to visitors in their various tourist activities wherever possible, depending on the length and nature of the trip (Meininger's International, 2021).

Wine tourism itineraries usually include visits to vineyards, wineries, and communes. Some are in-depth excursions to a single estate, while others are brief visits to various vineyards in the city or across the country. For tourists who want a more in-depth visit to one estate or visiting numerous estates in close vicinity, wine tour operators and wineries must urge guests to walk or ride rustic bicycles or e-bicycles. Not only would such a sustainable mode of transportation cut transportation-related GHG emissions, but it would also provide tourists a more immersive experience with nature than touring in vehicles such as cars and other public commercial transportation vehicles. It is worth noting that traveling by bicycle or foot rather than car is known as slow travel, and tourism that fosters such behavior is known as slow tourism. Slow tourism is gaining popularity in the wine tourism because of its environmental friendliness. The role of the national government would be essential in encouraging slow tourism. National governments must promote slow tourism by building infrastructure that allows tourists to rent bicycles or e-bikes. If a tour operator does not own a fleet of bicycles or e-bikes, they can either rent them or recommend locations where they can be rented.

When visiting estates that are not nearby but within the same city or across numerous cities, wineries and wine tour operators must encourage visitors to use intercity trains rather than going in gasoline-powered passenger vehicles like cars or choosing air travel. The majority of wine tourism destinations, including France, Italy, Spain, Portugal, and the United States, have a highly established intercity and cross-country train transport infrastructure. Wherever possible, using trains more frequently as opposed to flying or driving across the country is a more environmentally friendly option. Taking the train would greatly cut both road traffic and GHG emissions from transit. The author emphasizes that for journeys to vineyards, or wineries that involve air travel, travelers should consider booking flights with carbon-offset programs. Carbon offset programs can help offset a portion of an aircraft's carbon emissions during a specific flight by investing in carbon-cutting projects like reforestation and clean drinking water (Mowrey, 2021). Similarly, while visiting vineyards or wineries in less developed cities and urban districts with poor train connectivity, electric vehicles should be preferred over gasoline-powered motor vehicles (The Porto Protocol, 2021). The same advice should be given to non-residents traveling from other parts of the country, taking use of the chance to educate tourists on ecologically beneficial behaviors. Similarly, when wineries or visitors wish to hire commercial passenger vehicles for excursions, EVs should be the preferred choice. Gasoline-powered vehicles should only be utilized when absolutely essential. Using sustainable transport during wine tourism may have some cost repercussions for tourists, but it makes sense. A growing number of tourists, particularly Generation Z and millennials, are opting for environmentally friendly modes of transportation, such as electric vehicles or train travel, despite the additional expenditures. It has also been noticed that as people grow more environmentally conscious, they drive more environmentally friendly vehicles (Miller et al., 2022). A strategy like this would encourage tourists, who play an important role in the wine tourism industry, to adopt responsible behavior, lowering the sector's environmental impact. The author argues that if wine tour operators and wineries have the financial wherewithal, they should consider investing in electric (EV) or hybrid vehicles. This includes progressively transitioning from traditional gasoline-powered automobiles to e-vehicles, as well as investing in EV charging station infrastructure. Such a change would not only improve their company's reputation in the market as an environmentally friendly one, but it would also draw eco-aware tourists, expanding commercial potential. The high cost of EV ownership and maintenance has slowed their adoption in recent years, especially among smaller-scale wineries and wine tour operators. However, with future technological advancements in EV powered vehicles, such a barrier will be eliminated (Wood, 2022).

Many visitors choose luxury wine tourism packages with helicopter tours of vineyards and estates. Choosing drone with a virtual reality headset over helicopter rides in such occasions is a sustainable approach that wine tour operators and wineries can employ (Lee, n.d.). The combination of a drone and a virtual reality headset would provide tourists and visitors with real-time views of estates and vineyards, an environmentally responsible solution that would not compromise guest experience. Additionally, wine tour operators cater to tourists and visitors who choose quick wine tasting excursions lasting an hour or less. Sustainable practices for such events could include centralizing them in urban areas near hotels and restaurants, rather than going across the countryside from winery to winery. Such practices would cut not only GHG emissions from transportation, but also traffic congestion. These facilities can be found in many wine tourism destinations' regional towns. In California, for example, similar facilities are accessible in Napa, Sonoma, and Healdsburg (Meininger's International, 2021). The author recommends that wineries and wine tour operators provide tourists sustainability-focused tour packages. They must inform tourists about the environmentally friendly practices they use during the wine tour before to their arrival or during the trip booking process, such as through their website, email, or other promotional and marketing materials. This kind of approach would be just as beneficial as taking part in ecotourism initiatives (Marti & Durand, 2019).

Meetings Needs and Aspirations of Local Community

As previously said, social sustainability is one of the pillars of sustainable wine tourism. Winemakers and wine tour operators can accomplish social sustainability by preserving and promoting local culture, enabling community growth, and improving livelihoods by providing employment possibilities to neighboring communities (Szolnoki et al., 2022). To preserve and promote local culture and tradition, wine tour operators and wineries must work on community-based programs, one of which is educating visitors and tourists about local culture, its history, and its importance. An approach for educating tourists about local culture and tradition would be to display information folders that educate on local tradition, history, culture, and values in guest rooms, libraries, lounges, or any other place where tourists can spend time reading. Another good approach is to develop fee-based activities that allow tourists to visit villages and local people. Inviting elders from the local community to tell stories or share their experiences should also be the part of the program. It has been observed on multiple occasions that an influx of tourists can disrupt the natural serenity of the region, frustrating nearby communities (Cianga, 2017). This can lead to decreasing local community participation and poor tourist experiences, ultimately impacting the quality of community-based tourism experiences. It is crucial to remember that local leaders or the local administration must be included when organizing community-focused programs. They must be informed about the program, its goals, and its importance to the local community and the regional tourism industry as a whole. The proceeds from local tourism activities could be used to fund projects aimed at improving and developing the local community, as well as preserving local traditions and cultures. This involves cooperating with non-governmental organizations (NGOs) or local governments on community development initiatives such as the construction of schools, health care facilities, local water supply, and road construction (Eco Tourism Kenya, 2016). Furthermore, the author highlights that wine tour operators and local vineyards must invest a portion of their profits in environmental conservation or local development.

Improving Livelihood by Offering Employment Opportunities

Wineries and tour companies must also examine how to improve the quality of life for the surrounding communities (Yuki, 2021). This can be accomplished by offering competitively compensated job opportunities to members of the local community. Wine industry is overwhelmingly male-dominated and wineries and wine tour operators must take a strong stance in empowering women. Some of the best practices include providing job opportunities to women, particularly those from the local community, collaborating with women-owned wineries and bringing them clients, and highlighting their work on social media channels (Marsh, 2022). Such initiatives will encourage women to enter the wine sector, resulting in their empowerment. Minorities must be given employment chances as well. The author recommends hiring skilled workers and managers from the local community because it not only empowers them but also gives them the opportunity to earn and, ultimately, improve their standard of living. Ideally, 30% of your employees for skilled and unskilled work should be recruited locally (Eco Tourism Kenya, 2016). Locals, for example, could be hired as naturalists or tour guides because they are familiar with the area and have firsthand knowledge of the community, resulting in better information sharing with tourists and guests and, as a result, a better tourism experience. Because they depend on natural resources for a living, there is another good reason to take them into account for employment (Nagarjuna, 2015).

Other initiatives that can assist improve the livelihood of local communities include cooperating with local dance and entertainment groups to perform and allowing residents to sell handcrafted goods and artwork to resident tourists. Encourage tourists to participate in fishing or sailing sports, go on bike tours around town, stay in homestays or farm stays with local families, and purchase their homemade handcrafts or paintings to benefit the local community (Ingrassia, et al., 2022). Bringing tourists and guests to local vineyards in remote and underdeveloped areas would also be a step towards creating employment opportunities for local community members, raising earning potential, and, eventually, improving their quality of life. Efforts focused on the aforementioned goals would not only enhance their livelihood, but would also provide a competitive advantage to wine tour firms.

Promoting the Inclusion of Local Community Members

Locals from rural communities have been found to be unable to communicate with visiting tourists in English due to a lack of language proficiency (Kastenholz et al., 2023). This results in a lack of connection with tourists, resulting in a poor tourist experience, ultimately limiting progress towards social sustainability. Working with translators and investing in teaching English to community members, particularly young people, is one of the best methods to counteract this. This would not only increase their linguistic skills, but would also empower them. They should also be informed about what information to share with tourists and how to make them feel welcome. Local wine tour operators have also been found to lack significant wine tourism expertise, significantly impacting tourist experiences on wine tours. Wine tour operators can address this need by partnering with local wine tourism authorities to offer wine training courses (Berson, 2022). This would encourage young people to specialize in and pursue a career in wine tourism.

Collaborating with Sustainable Wineries

Another obvious action tour operator may take in sustainable wine tourism is to collaborate with and introduce clients to sustainable wineries, i.e., those who produce biodynamic or organic wines and incorporate sustainability into their core business operations (Filopoulos & Frittella, Designing sustainable and responsible wine tourism experiences, 2019; Trigo & Silva, 2022). Greater tourist visitation may boost consumer awareness of such wineries, resulting in greater demand for their wines. This type of collaboration would also help sustainably minded wineries stay in business, which is critical for the survival of sustainable wine tourism. It is critical to note that wineries rely on wine tour operators and travel agencies to communicate their sustainability values to tourists. However, being organic is not a prerequisite for forming a partnership. Collaboration with environmentally conscious wineries would also benefit wine tour operators because it would highlight the wine tour operators' commitment to sustainability. Furthermore, visitors and guests interested in sustainable wine tourism would be interested in learning about all of the partners with whom wine tour operators collaborate, and such collaboration would fulfill the expectations of tourists and guests of sustainable wine tourism, hence boosting tourist satisfaction. Such purposeful actions would also eventually help to change the destination's public image from one of mass tourism to one of eco-friendly, sustainable regions.

Promoting Local Tourism

Ensuring economic sustainability is a vital part of sustainable wine tourism (Szolnoki et al., 2022). Wine tour operators must work to promote local tourism in order to maintain their economic sustainability. There are many ways to accomplish this. One of the most straightforward approaches is to encourage tourists to visit tourist sites such as museums, art galleries, and sculpture gardens as well as engage in recreational activities such as shopping and sports. For example, tourists visiting France for the first time might have planned visiting certain popular and famous tourist attractions and missing few which are of tourism significance such as Monet's Garden in Giverny, Normandy, or Estérel Massif, a volcanic mountain in Golden Island—Var. Tourists visiting unplanned locations may generate business opportunities for hotels, tourist shops, restaurants, and other businesses that cater to tourists, as well as create employment opportunities for locals, which may aid in economic development while also encouraging entrepreneurship and new business ventures (Paulangelo, 2015). Studies have shown that wine tourism can significantly increase local GDPs (Tafel & Szolnoki, 2019). Restaurants and vendors who support local food producers and sustainable agriculture should also be suggested to tourists. Such an approach could improve future tourist influxes into the country. Increased tourism revenue can not only drive local governments to enhance local infrastructure, paving the way for additional tourism prospects, but it can also be reinvested in local economies, ranging from public services to education. It may also aid in the preservation of historic and cultural sites, so conserving local cultural traditions. Such efforts may have a big impact on the local economy (McClain, 2022).

Responsibly Conducting Wine Tasting Tours

Sustainable wine tourism requires wine tasting trips that are designed to protect participants' health and well-being while also delivering an enjoyable experience and promoting local heritage and culture (Filopoulos, 2021). Wineries and wine tour operators must ensure that participants at wine tasting events are served and sold wine responsibly. This might not only protect the participants' health and safety, but also ensure compliance with national alcohol laws. National authorities have devised norms and regulations to ensure the safe service of alcoholic beverages, and tourism operators are obligated to follow them. Integrating responsible practices into wine tasting tours requires work on multiple fronts. Minors (those under the legal drinking age) and high-risk individuals, such as pregnant and nursing women, are frequently present in wine tasting groups. To avoid injuring them, wine tour operators and wineries must refrain from providing wine to them. Additionally, no alcohol will be served to anyone under the legal drinking age during the wine tasting sessions. Another recommended practice for sustainable wine tasting tours is to limit the number of wines that a tourist can sample. Many participants may be disappointed since they would like to sample all of the wines. To avoid participant discontent, this should be explained at orientation meetings, the preliminary booking process, wine tour excursions, or prior to the wine tasting session. Posters, films, and messaging are all great ways to spread this information. When multiple wines must be served in a single session at a tasting event, participants must always be given metered pours (5 samples at 1 ounce each). This would ensure that they did not exceed the standard drink limit of 5 ounces of wine (Coravin, 2021).

Providing participants with snacks and water during tasting sessions is another noteworthy acceptable practice (Filopoulos & Frittella, 2019). Drinking water between sips would cleanse the palette and keep participants hydrated. Snacks, on the other hand, would assist balance alcohol consumption while also delivering a greater experience than wine alone. It is also crucial to provide participants with spittoons to guarantee that everyone can taste wine responsibly and adhere to responsible drinking guidelines. The author recommends that tourists be offered a wine tasting program that lasts 2–3 days. Tourists usually want to make the most of their wine tasting tours by trying as many wines as possible. However, such behavior puts individuals at risk of exceeding standard drinking guidelines. This can be decreased by offering attendees in-depth winemaking excursions spread out over a few days. This would allow participants to delve deeper into the process of producing quality wines while also reducing their drinking temptation. When attendees buy bottles of wine at wine tasting events, they should be reminded of drink they had during sampling sessions and urged to carry the bottle with them (Lavierge, 2018). Participants frequently drive from one winery to another, and being inebriated during the tour may put their safety at risk. Participants who intend to drive must be educated on the importance of not drinking any alcohol at all. This involves asking the group coordinator to choose the driver and urging the driver to abstain from drinking throughout the tour or, at the very least, to maintain a Blood Alcohol Concentration (BAC) below the legal limits. In such instances, a breathalyzer can be used to assess the driver's blood alcohol level prior to driving. A mechanism for rewarding drivers who refrain from drinking during the tour would also be beneficial. Wine tour operators and wineries could devise a program in which drivers are rewarded for not drinking alcohol during the wine tasting tour and for driving participants safely home. Tour providers and wineries might consider exchanging information about public transport and taxi services as necessary. These practices would not only teach participants about the dangers of drinking and driving, but would also protect participants' health and safety, giving guests more value and establishing better brand-customer connections (Filopoulos & Frittella, 2019).

Reducing Plastic and Paper Waste

Wine tour operators can implement many additional sustainable practices into their core operations, in addition to offering responsible wine tasting tours, to improve the sustainability of wine tourism activities. Implementing initiatives to reduce paper and plastic waste is one such practice. Some waste reduction initiatives include providing e-billing services, using rechargeable liquid soap in guest rooms, encouraging tourists to bring reusable carry bags, using collapsible bulk water containers during excursion tours, installing separate bins for different types of waste, and establishing a recycling contract with a local recycler. These environmentally focused initiatives limit waste entering landfills, resulting in less landfill-related pollution and lower greenhouse gas emissions into the environment (Green Dream Tours, n.d.). Another environmentally responsible strategy is to use locally produced goods rather than those brought in from other parts of the nation. Such measures could lower GHG emissions related to transportation and offer local businesses commercial opportunities. For example, buying fruits, vegetables, soap, and other locally produced things for resident guests.

The chapter's next section examines wineries' efforts to sustainability. Winemakers must take a holistic approach to winemaking that benefits the environment, conserves natural resources, protects wildlife, supports local communities, and safeguards consumer health.

Preserving Indigenous Grape Varieties

Endemic grapes should be prioritized in sustainable wine tourism. These grapes are geographically restricted and rarely grow in other parts of the country. When international grape varieties are preferred over endemic grapes in winemaking, the region's heritage is jeopardized. Winemakers must take advantage of the opportunity to make wines from indigenous grapes. Such steps would not only help to safeguard the region's historical grapes, but would also provide tourists with an authentic regional experience (Bulgaria Wine Tours, n.d.). The author highlights the significance of wine tour operators in supporting wineries in their efforts to protect indigenous grapes. Wine tour operators can assist by transporting tourists and visitors to wineries that produce wines from indigenous grapes. Such collaboration would benefit wineries, wine tour companies, the local population, and the region. It would provide wineries with business prospects and promotion, as well as help wine tour operators establish themselves as a sustainable firm, bring economic advantages to the local community, and contribute in the preservation of regional heritage and tradition. It is also prudent to educate tourists and guests on the importance of purchasing wines made from endemic grapes. Increased customer engagement with such wines would increase demand for such wines, encouraging wineries to continue with their endemic grape preservation efforts. Winemakers and wine tour operators can incorporate sustainability even further by pairing these wines with regional cuisine during tasting tours (Serra et al., 2021). It would provide a well-rounded gastronomic experience for tourists while also educating them about regional tradition and heritage, thereby improving their tourism experience.

Effectively Managing Winery Wastewater

Wineries' sustainability is determined by how they manage their wastewater (Saraiva et al., 2019). Wineries produce two forms of wastewater: cleaning wastewater (from washing barrels and other equipment) and sanitary wastewater (from restrooms). The discharge of their sanitary waste water, in particular, into sewage systems or bodies of surface water affects land, crops, and aquatic life while also contaminating surface and groundwater due to high levels of sugars and carbohydrates (Netsol Water, n.d.). Furthermore, growing wine grapes and creating wine require a significant amount of water, which is a limited natural resource. By using environmentally friendly practices such as recycling waste water, wineries can help to reduce water pollution while also alleviating the planet's water concerns by reducing demand for water supplies. Wineries must have an excellent water reclamation and recycling system in place in order to successfully utilize wastewater. The author highlights that there is no one-size-fits-all method for treating winery wastewater, and treatment choices are site specific. Before deciding on a treatment approach, many site-specific factors such as wastewater parameters, sludge management, available space, budget, and technical capability must be carefully evaluated.

Making lagoons or ponds (Oxidation, anaerobic, aerobic, shallow aerated) or installing mechanical systems are two proven methods for recycling and cleaning wastewater that wineries can consider (Laginestra, 2016). The author suggests that wineries might increase their recycled water supply even further by recycling rainwater for use in gardens and farms. Winery wastewater treatment can not only help wineries minimize their dependency on potable water for activities like gardening, toilets, and irrigation, cutting consumption and, ultimately, expenses, but it can also help wineries reduce environmental concerns associated with wastewater discharges. In addition to wastewater recycling, wineries must consider reducing their water use, which may be accomplished through a few simple procedures such as replacing water-saving spray guns with water-efficient versions, narrowing wine hoses, or repairing leaking hose (Garrison, 2022).

Transitioning to Solar Energy

The use of traditional fossil fuel-based energy (coal, oil, or gas) in wine production contributes significantly to the industry's carbon footprint (International Organisation of Vine & Wine, 2015). Wineries and other stakeholders in the wine tourism sector must use cleaner energy sources, such as solar energy, to reduce their carbon footprint (Carroquino et al., 2018). Solar energy has the potential to drastically reduce the environmental impact of wineries by eliminating reliance on traditional fossil fuel-based electricity. The advantages of solar energy are numerous. It can assist wineries in lowering energy consumption, power costs, and traditional electricity-related GHG emissions. Solar energy meets the energy needs of wineries since it can power production facilities, irrigate vines, and power offices, storage rooms, tourist housing, and tasting rooms (Chester Energy and Policy, n.d.). For example, the installation of a solar PV system has assisted De Bortoli Wines, Australia's sixth-largest winery, in saving thousands of dollars by significantly fulfilling their electricity needs (Abbott, 2016). Solar panels can also shield grapes from heat waves during hot summer days and insulate vines during chilly winters. Furthermore, their mobility allows more light to reach grape vines on cloudy days, resulting in improved growth. It has been found that grapevines covered by solar panels produce grapes with lower sugar content due to less light exposure, allowing for less alcohol during the fermenting process (Crellin, 2021). Over the last decade, a rising number of wineries have utilized solar energy and recognized the advantages of using solar energy on their production facilities. According to the author, while making investment decisions in solar systems, wineries should consider criteria such as the rate of return on investment, maintenance requirements, life expectancy, performance capability in local environmental circumstances, and the manufacturer's reputation. Another sustainable option that wineries can use to lessen their dependency on traditional fossil fuel-powered electricity is to build cellars into the landscape. Because natural limestone and subsurface spring water regulate air temperature, cellars constructed into the landscape function better in keeping cool temperatures (Senese & Esau, 2022).

Adopting Organic and Biodynamic Farming Practices

According to one study, a tonne of grapes grown using traditional agricultural practices emits 342 kg CO_2 eq due to the use of synthetic fertilizers and irrigation water supplied using traditional energy (Hefler & Kissinger, 2023). Furthermore, synthetic pesticides used in traditional grapevine farming injure the soil and leave harmful residues on grapes, putting consumer health at risk. Organic and biodynamic farming, both environmentally benign agricultural practices, can help wineries produce wine grapes without such consequences (Volanti et al., 2021). Over the last decade, organic grape planting and wine production has increased considerably. Organic farming enables winemakers to cultivate wine grapes without the use of synthetic fertilizers or pesticides, thereby saving the soil, the environment, and consumer health. It also improves soil fertility and production by providing nutrients and compost to the soil through crop and animal rotation practices. Better, nutrient-richer soil aids in the growth of higher-quality grapes, which in turn yields higher-quality wine with a greater nutrient profile that promotes consumer health. Organic farming also helps to reduce water use since organic farms have better soil structure, which allows for better water absorption and storage capacity, lowering water requirements (Leu, 2015).

Using biodynamic procedures to make wine is other environmentally friendly winemaking strategy that can help winemakers reduce their reliance on fertilizers and water consumption. In many aspects, biodynamic winemaking is similar to organic winemaking, including the use of manure and compost to boost soil fertility and the prohibition of synthetic fertilizers and pesticides. However, it differs from organic farming in several ways. For example, in biodynamic farming, only compost and manures produced on the farm are used to boost soil fertility, whereas in organic farming, compost and manures produced elsewhere can be used. Furthermore, biodynamic farming uses only natural pesticides such as neem and salt water, as opposed to organic farming, which allows for the use of certain synthetic pesticides in specific situations (Biodynamic Association, n.d.; Okafor, 2020). Regardless of whether organic or biodynamic winemaking practices are used, winemakers must inform tourists about the idea of sustainable winemaking procedures while conducting tourism activities. This will emphasize the importance of sustainability to them and motivate them to contribute to its realization, so advancing it (Singh, n.d.).

Implementing Barrel Sustainability

Winemakers should prioritize barrel sustainability since it is a key indicator of sustainability in the wine tourism sector. Over 250 different varieties of trees are utilized to produce oak barrels around the world, each with its own distinct flavor. Oak barrels have a finite lifespan, thus in order to make new oak barrels, hundreds of oak trees must be cut down over time, hurting the environment, wildlife, and climate. It is critical to recognize that harvesting mature trees is critical for forest regeneration and, consequently, forest health. Currently, the wine industry consumes more trees than it produces, thus impacting the ecosystem. To achieve sustainability in barrel use, winemakers must adopt specific tactics. They must use wood that has been sourced responsibly, such as wood with Programme for Endorsement of Forest Certification (PEFC) certification, wood with Sustainable Forestry Initiative (SFI) Certification, or wood derived from sustainably certified forests. These certifications ensure that the wood is sourced from previously stripped oak trees that have attained maturity. This type of endeavor would allow winemakers to receive the benefits of oak trees without endangering environment, wildlife, or the ecosystem. Wineries should also consider contributing a portion of their profits to forest conservation organizations in order to plant new oak trees. A way to contribute to the environment. Forests are a valuable natural resource, and wineries must commit to doing their share to protect them.

Another environmentally beneficial strategy for wineries would be to use 100% of the oak purchased because barrel manufacture generates a large quantity of wood waste. In wineries that have their own coopering, wood waste from barrel production could be fashioned into decor, furniture, or a flower pot. It can be used internally as a fuel or sold to other industries to make reclaimed wood items or barrel smoking wood for barbecuing. Purchasing oak wood from other parts of the country to construct barrels also contributes to GHG emissions. To remedy this, wineries may consider sustainably crafting barrels for their regional wines using wood from locally grown trees. Many winemakers prefer to utilize barrels made from trees grown locally. Vin Santo, an Italian dessert wine aged in local chestnut barrels, is one example. Making barrels from locally grown trees would not only serve to minimize deforestation and bring economic benefits to the community, but it would also allow tourists to experience terroir during regional wine tasting sessions (Senese & Esau, 2022). Worth the trip. Oak infusion solutions may also help wineries maintain sustainability by lowering their dependency on barrels. They can help winemakers reap the benefits of traditional oak barrels by recreating the distinct flavor and aroma of oak barrels (Petrozziello et al., 2020). Most notably, they are widely available and inexpensive, making them a feasible alternative. They are predicted to be in more demand in the future as winemakers become more concerned about environmental issues.

Using Environmentally Friendly Wine Packaging and Labels

Sustainable wine tourism comprises not just sustainable wine production but also the use of ecologically friendly packaging, such as labeling. The production of raw materials used in traditional labels and face stock, as well as their recycling, emits greenhouse gases that harm the environment, contributing to the wine industry's carbon footprint. Innovations and technological advances have opened the way for a variety of ecologically friendly labels that aim to reduce environmental impact. Winemakers now have a variety of environmentally friendly labels to choose from. This includes labels created from apple juice waste and 40% post-consumer recycled fibers, as well as marble-based film made from 80% calcium carbonate acquired from marble mining waste and up to 20% recycled high-density polyethylene (HDPE). Labels made of fiber paper made from 90% bagasse fiber obtained from sugar cane waste are also available as a sustainable option (Avery Dennison, n.d.; Juega, 2021). These environmentally friendly labels meet the demands of winemakers for printability, whiteness, and aesthetic appeal as well as the capacity to withstand colder temperatures. They are suitable for a number of applications, including still and sparkling wines (M_Use, n.d.).

Wineries must also consider the sustainability of their traditional PET and PP-based label release liners, which generate significant waste during the labeling process (Nedelcheva, 2019). Label release liners are essential for consistent and accurate label application. They are only recyclable in the presence of a formal waste collection system, and their recycling requires dedicated recycling infrastructure since the silicone that is a structural component of these liners must be removed during the recycling process. In the absence of organized trash collection system and specific recycling infrastructure, these release liners end up in landfills, increasing landfill-induced greenhouse gas emissions. To offset the environmental impact of these traditional label liners, wineries must collaborate with companies that offer collection and recycling programs for such liners, such as Avery Dennison and UPM Raflatac. These liners are recycled to create rPET (Recycled Polyethylene Terephthalate), which contains 15% recycled material (Juega, 2021). A sustainable approach to wine labels can surely contribute to the sustainability of the wine business and, eventually, its tourism sector.

Despite being indefinitely recyclable, traditional wine packaging, such as glass bottles, emits significant amounts of GHGs into the environment. Winemakers could explore utilizing more environmentally friendly packaging alternatives such as aluminum cans, stainless steel kegs, recycled plastic PET bottles, and Bag in a Box Packaging (BIB) for their wines. These packaging options are ideal for young wines. These packaging materials are more environmentally beneficial than glass bottles, even if they are not entirely sustainable and have their own environmental imprint in terms of production and recycling (Thompson-Witrick et al., 2021). According to the author, implementing environmentally friendly packaging is a strong approach to demonstrate and encourage sustainability.

Implementing Storytelling Practices

The author also emphasizes the importance of storytelling in achieving social sustainability. It forges a personal bond between the tourist, the winery, and wine tour business, encouraging tourists to work towards a more sustainable future (Santos et al., 2022). This includes informing tourists about a culturally significant animal featured in their symbol or naming their wine after a custom or culture. Such initiatives would expose tourists to local culture and strengthen ties, hence boosting tourists' experiences (Berson, 2022).

Getting feedback from tourists is just as important as organizing sustainable tourism activities. Inquiring into tourists' perceptions of sustainable practices used during tourism activities would help wineries and wine tour operators reevaluate their operations, opening up new opportunities to improve wine tourism sustainability. This approach must be viewed by wine tour operators and wineries as a moment of "truth." It would also allow for the enhancement of the tourist experience, hence increasing the brand-business relationship. The author emphasizes that the extent to which sustainable practices can be integrated into wine tourism operations is determined by the winemakers' and wine tour operators' philosophies as well as their financial capability, skill level, and availability of technology. Claiming to be sustainable without examining the nature of the organization's activities and underlying ideology may cause misunderstanding among tourists, resulting in a deterioration in business relations.

National Governments Promoting Sustainable Wine Tourism

The role of the national government would be critical in promoting sustainability in the wine tourism sector. National governments must build a framework for the development of truly sustainable resources and materials that the wine industry may use to accomplish their net zero aspirations. They must also prioritize the development of the nation's infrastructure for environmental sustainability, as well as implement legislation to incentivize sustainable business models in the wine industry (Mensah & Casadevall, 2019). Such strategic efforts will aid in the development of economically viable and readily available environmentally friendly materials and resources, as well as encourage industry participants to embrace sustainable business models, thereby accelerating sustainability adoption and making the sector sustainable. As businesses face multiple hurdles in transitioning to sustainable business models, it is critical to emphasize the relevance of national government support in assisting industry participants in adopting sustainable business models.

Conclusion

Adoption of sustainable practices in the wine tourism business is crucial to ensuring the industry's sustainability. Winemakers, wine tour operators, the local community, and tourists all benefit from sustainable wine tourism. It has the potential to improve visitor experiences, resulting in an increase in tourist arrivals and, ultimately, boosting the viability of wine regions worldwide, thereby protecting the wine tourism sector. Additionally, it supports regional and national economic development while preserving cultural assets and raising the living standards of surrounding communities (Sosa, 2022).

In this chapter, the author took a straightforward, comprehensive, and elaborative approach to identifying best practices to enhance sustainability in the wine tourism industry. An examination of previous research on the subject revealed a patchwork of thorough information on the subject. The chapter adds to the subject by providing detailed and informative information and insights. Although the author claims that this chapter is insufficient to make an impact, it is organized in such a way that understanding of the issue is more likely, promoting the adoption of sustainability initiatives. This chapter could be a useful starting point for sustainability managers and strategists at wineries and wine tour companies. Overall, the author believes that the chapter will increase the possibility that winemakers and wine tour operators in the wine tourism sector would recognize the importance of making the industry more sustainable. The author urges winemakers and wine tour operators to include sustainability into their core business functions as much as feasible. Profits must not come at the expense of sustainability. Winemakers and wine tour operators have a unique opportunity and responsibility to foster sustainability in the wine tourism industry by demonstrating their commitment to sustainability in their various fields of work and educating tourists on the importance of sustainability (Szolnoki et al., 2022). Leading by example would also encourage tourists to join the sustainability movement, so spreading it globally.

References

Abbott, T. (2016). *Winery saves following solar panel install.* Wine Australia.

Amarando, M., Assenov, L., & Visuthismajam, P. (2019). Sustainable wine tourism and vineyards' environmental consciousness in Thailand. *African Journal of Hospitality, Tourism and Leisure, 8*(3).

Avery Dennison. (n.d.). *Earth friendly wine labels with extra shelf appeal*. Avery Dennison. Retrieved February 23, 2023, from https://label.averydennison.com/eu/en/home/products/wine-labels/sustainable-packaging.html

Berson, M. (2022). *The role of wine tour operators in the development of sustainable tourism in Crete*. University of Porto.

Biodynamic Association. (n.d.). *Biodynamic principles and practices*. Biodynamic Association. Retrieved March 17, 2023, from https://www.biodynamics.com/biodynamic-principles-and-practices#:~:text=Biodynamics%20Generates%20On%2DFarm%20Fertility&text=Composting%20brings%20animal%20manures%2C%20plant,manures%20that%20nurture%20the%20soil.

Bulgaria Wine Tours. (n.d.). *Melnik & Struma Valley: endemic grape varieties, culture and SPA*. Bulgaria Wine Tours. Retrieved January 31, 2023, from https://bulgariawinetours.com/melnik-struma-valley-endemic-grape-varieties-nature-spa/

Carroquino, J., Casarejos, N., & Gargallo, P. (2018). Introducing renewable energy in vineyards and agricultural machinery: A way to reduce emissions and provide sustainability. *Wine Studies, 6*(1). https://doi.org/10.4081/ws.2017.6975

Chester Energy and Policy. (n.d.). *Solar power and wineries: A match made in Heaven…and California*. Chester Energy and Policy. Retrieved January 14, 2023, from https://www.chesterenergyandpolicy.com/blog/solar-power-wineries-california

Chrobak, U., & Zimmer, K. (2022, September 12). *How climate change is tweaking the taste of wine*. BBC. Retrieved January 1, 2023, from https://www.bbc.com/future/article/20220825-how-climate-change-affects-wine

Cianga, N. (2017, May). The impact of tourism activities. A point of view. *Risks and Catastrophes Journal, 20*(1), 25–40. https://doi.org/10.24193/RCJ2017_02

Coravin. (2021, April 12). *Wine in Moderation: Serving and drinking responsibly*. Retrieved March 9, 2023, from https://www.coravin.com/community/wine-in-moderation-drinking-responsibly

Crellin, F. (2021, October 8). *Solar panels help French winemaker keep climate change at bay*. Reuters. Retrieved February 1, 2023, from https://www.reuters.com/business/environment/solar-panels-help-french-winemaker-keep-climate-change-bay-2021-10-04/

Dang, B. (2017, September). *Travelers' awareness and attitude towards environmentally sustainable tourism in Helsinki*. Haaga Helia University of Applied Sciences.

Eco Tourism Kenya. (2016). *How to engage the local community in your tourism business*. Eco Tourism Kenya.

Filopoulos, S. (2021). Sustainable business models of wine tourism to promote responsible consumption. In *Sustainable and innovative wine tourism. Success models from all around the world*. Cajamar Caja Rural.

Filopoulos, S., & Frittella, N. (2019). *Designing sustainable and responsible wine tourism experiences*. 41st World Congress of Vine and Wine. 12. BIO Web of Conferences.

Garrison, B. (2022, July 8). Diving into winery water usage and treatment. The *Grapevine Magazine*. Retrieved February 17, 2023, from https://thegrapevinemagazine.net/2022/07/diving-into-winery-water-usage-and-treatment/

Green Dream Tours. (n.d.). *Our commitment to sustainable tourism*. Retrieved March 7, 2023, from https://greendreamtours.com/sustainability/

Hefler, Y., & Kissinger, M. (2023). Grape wine cultivation carbon footprint: Embracing a life cycle approach across climatic zones. *Agriculture, 13*(2). https://doi.org/10.3390/agriculture13020303

Ingrassia, M., Altamore, L., Bellia, C., Grasso, G., Silva, P., Bacarella, S., Columba, P., & Chironi, S. (2022). Visitor's Motivational Framework and Wine Routes' Contribution to Sustainable Agriculture and Tourism. *Sustainability, 14*(19). https://doi.org/10.3390/su141912082

International Energy Agency. (2022). *Improving the sustainability of passenger and freight transport*. Paris: International Energy Agency. Retrieved February 6, 2023, from https://www.iea.org/reports/transport

International Organisation of Vine and Wine. (2015). *Greenhouse gases accounting in the vine and wine sector—Recognised gases and inventory of emissions and sequestrations*. International Organisation of Vine and Wine.

Jones, G., & Webb, L. (2010). Climate change, viticulture, and wine: Challenges and opportunities. *Journal of Wine Research, 21*(2), 103–106. https://doi.org/10.1080/09571264.2010.530091

Juega, M. (2021, April 29). *Can we make wine barrels more sustainable?* (S. W. Everyone, Producer). Sustainable Wine for Everyone. Retrieved May 2,

2023, from https://sustainablewineforeveryone.com/can-we-make-wine-barrels-more-sustainable/

Karagiannis, D., & Metaxas, T. (2020). Sustainable wine tourism development: Case studies from the Greek region of Peloponnese. *Sustainability, 12*(12). https://doi.org/10.3390/su12125223

Kastenholz, E., Paco, A., & Nave, N. (2023, January). Wine tourism in rural areas—Hopes and fears amongst local residents. *Worldwide Hospitality and Tourism Themes, 15*(1), 29–40. https://doi.org/10.1108/WHATT-08-2022-0095

Laginestra, M. (2016, January–February). *Winery wastewater treatment and attaining sustainability*. https://www.ghd.com/en/expertise/resources/PDF/Winery_wastewater_treatment.pdf

Lavierge. (2018, December). *4 Ways to practice responsible drinking*. Retrieved March 21, 2023, from https://lavierge.co.za/4-ways-to-practice-responsible-drinking/

Lee, I. (n.d.). *How Drones are helping with sustainability in the wine industry*. Drone Blog. Retrieved March 2, 2023, from https://www.droneblog.com/how-drones-are-helping-with-sustainability-in-the-wine-industry/

Leu, A. (2015). Increasing soil matter through organic agriculture. *Acres U.S.A. Magazine*. Eco Farming Daily. https://www.ecofarmingdaily.com/build-soil/increasing-soil-organic-matter-through-organic-agriculture/

M_Use. (n.d.). *A sustainable label for a premium vintage*. M_Use. Retrieved April 12, 2023, from https://www.my-muse.com/en/home/insights/recycled-wine-label-range.html

Marsh, J. (2022, January 13). *How community-based tourism benefits local communities*. Sustainable Tourism Consultancy. Retrieved April 1, 2023, from https://www.seagoinggreen.org/blog/how-community-based-tourism-benefits-local-communities

Marti, C., & Durand, S. (2019). *How to encourage sustainable tourism*. Horwath—Hotel, Tourism and Leisure. https://cdn.horwathhtl.com/wp-content/uploads/sites/2/2019/05/Industry-Report_Sustainable-Tourism.pdf

McClain, J. (2022, November 30). *What is wine tourism and how does it benefit the local economy*. Tourism Review News. Retrieved April 7, 2023, from https://www.tourism-review.com/wine-tourism-benefits-local-economies-news12827

Meininger's International. (2021, November 5th). *Sustainable wine tourism: the next big thing?* Retrieved March 3, 2023, from https://www.meiningers-international.com/wine/marketing-wine-tourism/sustainable-wine-tourism-next-big-thing

Mensah, J., & Casadevall, S. (2019). Sustainable development: Meaning, history, principles, pillars, and implications for human action: Literature review. *Cogent Social Sciences, 5*(1). https://doi.org/10.1080/23311886.2019.1653531

Miller, R., Cardell, M., Batra, G., & Goel, A. (2022, May 23). *Why consumers are charging toward electric vehicles.* Ernst & Young. Retrieved January 17, 2023, from https://www.ey.com/en_it/automotive-transportation/mobility-consumer-index-wave-3

Montella, M. (2017). Wine tourism and sustainability: A review. *Sustainability, 9*(1). https://doi.org/10.3390/su9010113

Mowrey, L. (2021, June 30). *How to make wine travel more sustainable.* Wine Mag. Retrieved January 14, 2023, from https://www.winemag.com/2021/06/30/sustainable-wine-travel/

Nagarjuna, G. (2015). Local community involvement in tourism: A content analysis of websites of wildlife resorts. *Atna Journal of Tourism Studies, 10*(1). https://doi.org/10.12727/ajts.13.2

Nedelcheva, M. (2019). *Label liners: meeting the sustainability challenge.* Avery Dennison. https://label.averydennison.com/content/dam/averydennison/lpm-responsive/europe/english/images/products/film/sustainability-clear-intenst/white-paper-on-liner-sustainability.pdf

Netsol Water. (n.d.). *How to treat waste water in wineries or liquor producing units?* Netsol Water. Retrieved January 16, 2023, from https://www.netsolwater.com/waste-water-treatment-solutions-for-wineries-or-liquor-producing-units.php?blog=2064

Okafor, J. (2020, June 12). *What is biodynamic farming and why is it important?* TRVST. Retrieved May 1, 2023, from https://www.trvst.world/environment/what-is-biodynamic-farming-and-why-is-it-important/

Paulangelo, N. (2015, January 8th). *12 Best wine practices in global wine tourism.* Nico Paulangelo. Retrieved March 18, 2023, from https://nicopaulangelo.wordpress.com/2015/01/08/12-best-practices-in-global-wine-tourism/

Petrozziello, M., Nardi, T., Asproudi, A., Cravero, M., & Bonello, F. (2020). Chemistry and Technology of Wine Aging with Oak Chips. In F. Cosme, F. Nunes, & L. Ribeiro, *Winemaking—Stabilization, aging chemistry and biochemistry* (p. 256). https://doi.org/10.5772/intechopen.93529

Pomarici, E., & Vecchio, R. (2014). Millennial generation attitudes to sustainable wine: An exploratory study on Italian consumers. *Journal of Cleaner Production, 66*(1), 537–545. https://doi.org/10.1016/j.jclepro.2013.10.058

Santos, V., Dias, A., Ramos, P., Madeira, A., Sousa, B., & Santarem, I. (2022). The influence of wine storytelling on the global wine tourism experience. *Wine Economics and Policy, Just Accepted*. https://doi.org/10.36253/wep-11454

Saraiva, A., Rodrigues, G., Mamede, H., Silvestre, J., Dias, I., Feliciano, M., Silva, O., & Oliveira, M. (2019). The impact of the winery's wastewater treatment system on the winery water footprint. *Water Science and Technology, 80*(10), 1823–1831.https://doi.org/10.2166/wst.2019.432

Senese, D., & Esau, D. (2022). Sustainable transformative wine tourism: Applications in experiential educational tourism. In *The future of luxury brands* (pp. 231–250). De Gruyter. https://www.degruyter.com/document/doi/10.1515/9783110732757-012/html?lang=en

Serra, M., Antonio, N., Henriques, C., & Afonso, C. (2021). Promoting sustainability through regional food and wine pairing. *Sustainability, 13*(24). https://doi.org/10.3390/su132413759

Singh, V. (n.d.). *How organic wine benefits you and the environment*. The Momentum. Retrieved February 4, 2023, from https://www.themomentum.com/articles/how-organic-wine-benefits-you-and-the-environment

Sosa, B. (2022, June 23). *Wine tourism: Making journeys through vineyards more sustainable*. Preferred by Nature. Retrieved April 11, 2023, from https://preferredbynature.org/newsroom/wine-tourism-making-journeys-through-vineyards-more-sustainable

Stefanica, M. (2017, July). Environmental impact of transportation in the tourism industry-dimensions and actions. *Studies and Scientific Researches Economics Edition, 25*. https://doi.org/10.29358/sceco.v0i25.377

Sun, Y., & Drakeman, D. (2021, August 9). *Vineyard tourism is a big source of carbon emissions. Want to help? Then buy more wine*. University of Queensland, Australia. Retrieved January 14, 2023, from https://business.uq.edu.au/article/2021/08/vineyard-tourism-big-source-carbon-emissions-want-help-then-buy-more%C2%A0wine

Szolnoki, G., Nilsson, C., Tafel, M., Ridoff, N., & Stelter, A. (2022). *Sustainable wine tourism—Global survey.* Geisenheim University and Wine-Tourism.com. Geisenheim University and WineTourism.com. https://www.winetourism.com/files/2022/01/Sustainable_wine_tourism_Report_GeisenheimUni_WineTourismcom.pdf

Tafel, M., & Szolnoki, G. (2019). Estimating the economic impact of tourism in German wine regions. *International Journal of Tourism Research, 22*, 788–799. https://onlinelibrary.wiley.com/doi/epdf/10.1002/jtr.2380

The Porto Protocol. (2021, December 13). *Sustainable wine tourism—What is it? How can we achieve it?* Retrieved January 21, 2023, from https://www.portoprotocol.com/sustainable-wine-tourism-what-is-it-how-can-we-achieve-it/

Thompson-Witrick, K. P., Nemenyi, J., & Budner, D. (2021, June 7th). The Impact Packaging Type Has on the Flavor of Wine. *Beverages, 7*(2). https://doi.org/10.3390/beverages7020036

Trigo, A., & Silva, P. (2022). Sustainable development directions for wine tourism in douro wine region, Portugal. *Sustainability, 14*(7). https://doi.org/10.3390/su14073949

Volanti, M., Martinez, C., Baeza, E., Vassura, I., & Passarini, F. (2021). Environmental sustainability assessment of organic vineyard practices from a life cycle perspective. *International Journal of Environmental Science and Technology, 19*, 4645–4658. https://doi.org/10.1007/s13762-021-03688-2

Wood, J. (2022, October 26). *Electric vehicles: The 3 main factors holding back sales*. World Economic Forum. Retrieved February 5, 2023, from https://www.weforum.org/agenda/2022/10/ev-sales-charging-infrastructure-transport-sector-sustainable/

Yuki. (2021, August 11). Opportunities for sustainability in wine tourism. *Borgen Magazine*. Retrieved March 9, 2023, from https://www.borgenmagazine.com/wine-tourism/

Wine Tourism, the Business of Wine, and the Impact of the Environöment and Sustainability

Donna Lee Rosen and Doris Miculan Bradley

Sustainability and Wine Tourism

According to the United Nations, climate change has been named the defining issue of the twenty-first century. The impact of global warming has consequences for current and future global humanity. Immediate changes to good consumption and drinking habits are needed if humankind is to achieve the goals identified in the Paris Agreement of keeping climate change under 2 °C by 2050.

D. L. Rosen
Royal Roads University, Victoria, BC, Canada
e-mail: dleerosen@georgebrown.ca

D. Miculan Bradley
George Brown College, Toronto, ON, Canada

Present Address:
D. L. Rosen
SHTM, George Brown College, Toronto, ON, Canada
e-mail: dorisbradley@georgebrown.ca

J. Martínez-Falcó et al. (eds.), *Wine Tourism and Sustainability*,
https://doi.org/10.1007/978-3-031-48937-2_6

While viticulture, viniculture, and climate change may seem like independent issues, the way grapes are selected for wine production, irrigate vineyards, bottle wines, and move wines globally attest to climate change and capitalistic planting for awarded wines. One of the most impactful things we can do as individuals is to ignore the chase for gold medals and support producers that are leading the globe's best environmental practices in the business of grape growing, wine production, and global supply chain management.

With sustainable practices in the business of grape growing and winemaking, the industry needs to be positioned to make immediate changes to its livelihood. The negative impact of synthetic fertilizers and pesticides, water-dependent species, fossil fuel consumption, and a poor circular economy play into the issue of why the wine industry finds itself at this juncture. As some scholars suggest that sustainable farming is a journey, grape growers and winemakers should be addressing the change as a race towards the survivability of planet earth and the future of human beings.

This chapter will be dedicated to the change agents leading the best practices in the commerce of grapes, wine, and transportation and how consumers can best prepare themselves to purchase wines and responsibly visit wineries.

Wine Tourism Defined

Tourism is composed of many niche areas that appeal to a plethora of people globally. While traveling is the act of getting from one place to another, tourism is there and consuming the riches each destination has to offer. The learning process, enjoyment, consumption, and appreciation of wine find its way as one such destination offering. The UNWTO defines tourism as "Tourism is a social, cultural and economic phenomenon which entails the movement of people to countries or places outside their usual environment for personal or business/professional purposes" (UNWTO—Gastronomy).

As a tourism sector, wine tourism is estimated to grow from about 10% in 2020 to almost 30% of the worldwide tourism market share by 2030 (Statistica.com market size 2021).

This is interesting when one considers that wine consumption trends have been heading down over the past years (Statistica.com, 2021). Consideration to the following definitions of wine tourism must be included as we continue to explore the sector and its future.

Wine tourism.com sees wine tourism as any activity that allows visitors to experience the winery, the vineyards, or the surrounding wine region (winetourism.com, 2021). Wine tourism has been defined by Hall in 2002 as "visitation to vineyards, wineries, wine festivals and wine shows for grape wine tasting and/or experiencing the attributes of a grape wine region" (Gomez et al., 2015). There have also been words written by Charters and Ali-Knight offering that wine tourism experiences includes many contents such as wine, food, art, tasting experience, cellar door activities, cultural tours…" (Guzel et al., 2021). Byrd et al. (2016) more recently has been described as "travels to wineries and what the wind tourist are experiencing at these destinations" (Santos, 2019). In summation, the wine tour comprises many components of tourism as a whole, with a main purpose for the destination decision as that of the wine production at that destination.

The UNWTO notes too that Wine tourism, also known as Enotourism is a sub-type of Gastronomy Tourism, refers to tourism whose purpose is visiting vineyards, wineries, tasting, consuming, and/or purchasing wine, often at or near the source (UNWTO—Gastronomy).

Wine enthusiasts choose their destination around wine regions, varietals, and appellations. The goal is to learn more and taste what they cannot find at home or to further delve into a region or appellation they have a desire for—based on taste and what they can afford! After those issues are dealt with in the first step of the path to purchase, accessibility/offerings of vineyards for tourism and seasonality are key (www.worldsbestwines.eu).

According to Santos et al. wine tourism is an experience that seems to have three key dimensions as follows: situation, knowledge decoding, and emotional response. Therefore, it was summarized as a response where the consumer decodes cues and gains some extra knowledge that can provide an enjoyable memorable or unique emotional response that can be transmitted to others."

Finally, the UNWTO notes that "wine tourism is intimately related to the identity of destinations and comprises cultural, economic and historical values" (UNWTO Wine Tourism). As a growing sector of tourism, it can be surmised that wine tourism is not only for wine enthusiasts, but for those interested in gaining knowledge about the history, culture, and experience surrounding visiting of wine destinations to share emotional responses being transmitted to and from others. The experience of visiting destinations that have a wine tourism component is something that does not exist in a vacuum. Culture, art, history, experience, and gaining knowledge all come together to benefit the economy and develop a sustainable tourism future.

The Impact of Wine Tourism

While enotourism has been a growing niche within the tourism industry, it is not a new phenomenon. For centuries people have been traveling to either purchase, taste, and experience, participate in product creation, and learn about wines and winemaking. In fact, many have opened a bottle just to remind themselves of a destination from which it came.

The origin of wine tourism in traditionally wine-producing countries was otherwise known as the "wine road" (Encyclopedia MDPI). Wine tourism has been defined as "one who travels to wineries and what the wine tourists are experiencing at these destinations" (Byrd et al., 2016, p. 20).

By the twenty-first century wine and food tourism has created a definitive niche among tourists globally (Sharples, 2002). Wine and food have become one of the travel motives for tourists around the world (Sharples, 2002).

Travel and Tourism contributed 5.81 trillion USD in 2021 to the global GDP, which was a one trillion USD increase from 2020 (statista.com).

While these numbers are down from the years prior to the pandemic (2020–2022) they are a significant contributing factor to the GDP and any destination that produces tourism as an export. This increase mid-pandemic is significant and only reinforces the strength of tourism globally and the value it has to the global economy.

France continues to be the country with the highest number of international tourist arrivals worldwide in 2021. Respectively, Mexico and Spain ranked second and third (statista.com). Not surprisingly the top three wine-producing countries in 2021 were Italy, Spain, and France in that order (World Population Review). While it is hard to draw a correlation between tourism to the destination and their wine production without specific tracking as to the motive of the tourists, it is safe to consider that interest in the wines of that region may be a strong contributing factor to the positive experiences of many of the tourists traveling to those destinations.

At this time data is not collected with any systematic method in 75% of countries, and there is not national programmed aimed at improving data collection 50% of countries regarding wine tourism as there is a lack of resources, lack of motivations for wineries to provide such data and any data collection has been done randomly and therefore cannot be conflated with any type of comparability (UNWTO).

Despite this, the wine tourism market worldwide was estimated to be worth around 8.7 billion US dollars in 2020, a year in which the entire global tourism industry was hit hard by the coronavirus (COVID-19) pandemic. As forecast, the ecotourism's market size was expected to reach nearly 29.6 billion euros in 2030 (UNWTO). There is a challenge however as based on the 2018 Global status report on alcohol and health published by the World Health Organisation (WHO), "*the percentage of countries with a written national alcohol policy steadily increased from 2008, and many countries have revised their policies since the Global strategy to reduce the harmful use of alcohol was released*" leading to a stricter regulatory environment in many countries (WHO, 2018).

Culturally, wine has been part of the lifestyles of many cultures globally, some daily rituals, some as celebratory assets and many for all occasions. There is a social, cultural, agricultural, environmental, and economic contribution to many destinations because of wine consumption. Despite the lack of resources to track data in understanding wine tourism trends along with the new alcohol policies, wine export reports support the consideration that wine and gastronomic tourism continues to be of increased interest to many. France, Italy, and Spain are the top three exporters of wine, from 2020 to 2021 export trends of wine increased by 4% globally.

It is valuable to note that consumption trends have decreased by 1% between 2021 and 2022, as millennials are switching to non-alcoholic beverages. According to Wine News Jan 2022 many are moderating their wine consumption by consuming water between glasses (Campbell & Bob, 2022).

The wide ecosystem that surrounds wine consumption, development, curation, and tradition remains strong as a tourist experience. Supporting the sustainability of wine regions is important to safeguard this valuable cultural heritage and the positive environmental impact to the destinations they belong to (Wine Intelligence report).

As noted by Russo et al., new trends and technologies, new growing markets and the increasing restrictive policies and regulation, create a challenging environment for the wine business, which could in turn foster innovation and sustainable business models (Russo et al., 2015). Therefore, this "challenge" is only a catalyst to speed up the road to the development of sustainable wineries and therefor supporting sustainable tourism demands. The value of vineyards reaches beyond the scope of culture, heritage, and culinary tastes; it is an integral part of sustaining the environment and working towards achieving the Sustainable Development Goals (SDGs) as per the United Nations (UN).

In 2016, the United Nations World Tourism Organisation (UNWTO) organized the 1st UNWTO Global Conference on Wine Tourism in Georgia. As a result of this conference, there a publication emerged titled the Georgia Declaration on Wine Tourism. This sets the framework for Wine Tourism as part of Gastronomy and Cultural Tourism. Wine tourism is now recognized as a key element for both emerging and mature tourism destinations while fostering sustainable tourism development (UN.org). This conference is an annual event that met again in 2017 and 2018 fostering further focus on "Wine Tourism as a Tool for Rural Development" linking directly to the outcomes being sought by the 2030 Agenda for Sustainable Development and the 17 Sustainable Development Goals (SDGs) (UN.org). The 2018 meeting further highlighted the need for partnership and innovative models of collaboration as well as local community engagement and empowerment (UN.org).

A very timely recognition as until recently "natural" wine was a relatively new concept in the wine world and one which wine consumers appear to have increased interest (Wine Intelligence, 2022). Natural wine is considered a subset of organic wine and theoretically implies that there is less intervention from the winemaker in the production of this product (Makish, 2021) therefore also reducing the carbon emissions? Significantly, between 2004 and 2019 the sales of organic wines increase by 28%. This increase in popularity for the organic wine product demonstrates a demand for further consideration of wine tourism as part of sustainable development again promoting sustainable wine tourisms' positive impact on tourism that supports (see Fig. 1).

As a result, increased demand for tourism will result in increased demand for wine production and therefore increased development of organic wine offerings. The value of sustaining sustainable wine tourism is emphasized by the increased desire towards responsible travel

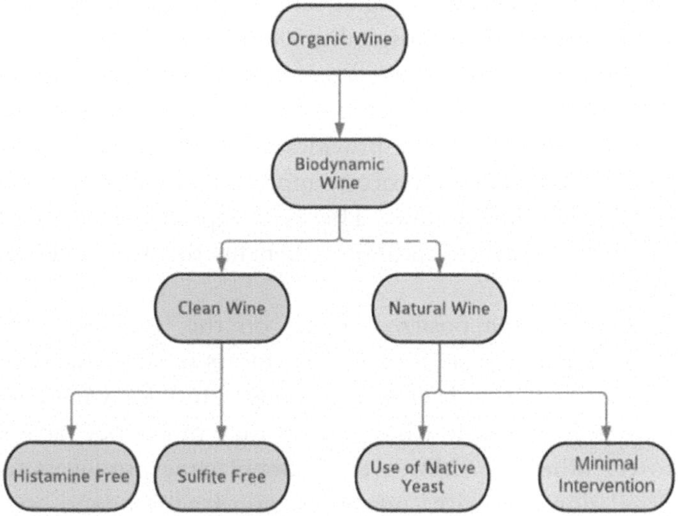

Fig. 1 Classification of organic wine (Makish, 2021)

supporting the achievement of the United Nations Sustainable Development Goals. Support of organic wine tourism increases productivity of the land, offers employment opportunity, supports responsible consumption and production, and life on land. The organic wine production methods eventually lead to land and water that is no longer impacted by harmful fertilizers and other substances used to produce high intervention products, leaving behind clean fertile land and clean water.

Wine Tourism for the Wine Industry

Wine tourism has a significant impact on overall business, particularly in the wine and hospitality sectors. Wine tourism provides an opportunity for wine producers to showcase their products, establish direct relationships with consumers, and diversify their revenue streams. It also supports the hospitality industry by providing opportunities for businesses to offer services to visitors, generating employment opportunities, and promoting regional development.

Enotourism, oenotourism, otherwise known as wine tourism, is a growing industry that has a significant impact on overall business, particularly in the wine and hospitality sectors.

Wine tourism refers to the practice of visiting wine-producing regions and participating in wine-related activities, such as tastings, tours, and vineyard walks. The experience provides an opportunity for tourists to learn about the winemaking process, appreciate the beauty of the vineyards, and taste the final product. This essay explores the impact of wine tourism on overall business and how it benefits the wine and hospitality industries.

Wine tourism has a positive impact on the wine industry. Wine tourism provides an avenue for wine producers to showcase their products to a wider audience. It creates an opportunity for wine enthusiasts to visit vineyards and taste wines, which they might not have had the chance to try otherwise. Moreover, it allows wine producers to establish direct relationships with consumers, gain feedback, and understand their preferences. This feedback can help producers improve their products and tailor their offerings to meet the evolving tastes of consumers.

Wine tourism also helps wine producers to diversify their revenue streams by offering wine-related activities, such as vineyard tours, tastings, and wine-pairing dinners.

The hospitality industry has always been a benefactor of wine tourism. Wine tourism attracts visitors who are interested in exploring the wine regions and experiencing the local culture. This demand provides opportunities for hospitality businesses, such as hotels, restaurants, and tour operators, to offer services to visitors. The increased demand for these services generates employment opportunities and supports the local economy. Moreover, wine tourism can promote regional development by encouraging the establishment of new businesses, such as artisanal food producers, boutique hotels, and transportation services.

The overall economy will be impacted in a positive manner. Wine tourism can contribute to the growth of the local economy by increasing employment opportunities, generating tax revenue, and stimulating the development of new businesses. Furthermore, wine tourism can promote international trade by creating demand for wine products from different regions. This demand can result in increased exports, which can further contribute to the growth of the economy.

However, wine tourism can also have negative impacts on overall business. For example, wine tourism can put pressure on local resources, such as water, energy, and land. The increased traffic can also lead to congestion, noise, and pollution, which can negatively impact the quality of life for local residents. Wine tourism can also increase the risk of alcohol-related incidents, such as drink driving, which can lead to safety concerns.

It is important to note that it is imperative to manage the negative impacts of wine tourism to ensure sustainable development of the industry.

Sustainability and Wine Tourism

Defining sustainable wine tourism may be considered through various lenses. One is to focus on the sustainability of the business itself and the other is to focus on the process of wine production being environmentally sustainable.

According to the Brundtland Commission, as organized by the United Nations in 1983, "Sustainability is a holistic approach that considers ecological, social and economic dimensions, recognizing that all must be considered together to find lasting prosperity." Breaking down that definition, it has been determined that sustainability has three pillars. Firstly, an economic pillar, secondly, an environmental pillar and finally, a societal pillar. In order to focus on the economic sustainability of the environmental pillar, we must define what it means for wine production to be environmentally sustainable and secondly what the economics of that sustainable wine production process is.

A widely accepted definition of sustainability that has been adopted as a result of the Brundtland Commission is that "development that meets the needs of the present without compromising the ability of future generations to meet their own needs" (United Nations Academic Impact, 1989).

Economic sustainability then is described as; "Human communities across the globe being able to maintain their independence and have access to the resources that they require, financial and other, to meet their needs" (University of Alberta, 2019). According to the University of Mary Washington, Office of Sustainability, "Economic sustainability refers to practices that support long-term economic growth without negatively impacting social, environmental, and cultural aspects of the community" (UNAI, 2019). Given these two widely accepted definitions, we can now apply that concept to the parameters of environmentally sustainable wine production.

The ability to determine what is required for wine tourism to survive perpetually and independently without negative impact on other areas will be established. A positive outcome would then further support the initiatives driving the industry to further become sustainable in its quest to become environmentally sustainable.

As recently as September 2022, the United Nations World Tourism Organization (UNWTO) held a global wine tourism conference focused specifically on innovation, sustainability, and creativity for the wine industry. This highlights the relevance and significance of understanding, considering, and applying parameter of the sustainability practices for sustainable wine tourism practices as immediate.

Demand for Environmentally Sustainable Wines

The demand for environmentally sustainable products, including wine, is on the rise, as more global consumers are becoming conscious of their impact on the environment and are looking for products that align with their values.

Sustainable wine is a broad term that encompasses several practices, including organic and biodynamic farming, renewable energy, carbon neutrality, and sustainable packaging.

Many consumers view organic or biodynamic wines as healthier and better for their overall well-being, as they are produced without synthetic pesticides, fertilizers, or other harmful chemicals. Furthermore, consumers are becoming more aware of the impact of their purchasing decisions on the environment and are seeking out products that minimize their carbon footprint and promote sustainability.

Some consumers are looking for products that support fair labor practices and ethical sourcing, and environmentally sustainable wines often promote these values. Sustainable wine producers often focus on quality and terroir, which can lead to unique and delicious wines that stand out from mass-produced wines. As a result, many wine producers are investing in sustainable practices and labeling their products as environmentally friendly or sustainable to meet the growing demand from consumers. Overall, the trend towards environmentally sustainable wines is likely to continue as consumers become increasingly conscious of their impact on the environment and demand more sustainable products.

Operating a Sustainable Wine Tourism Business

Running a successful wine tourism business requires a combination of factors, including a deep knowledge of the local wine industry, an understanding of hospitality and customer service, a passion for wine, and a commitment to sustainable practices. By focusing on these factors and creating a memorable wine tourism experience, businesses can attract visitors and contribute to the growth of the wine tourism industry.

Wine tourism is a growing industry that offers visitors a chance to experience the culture, history, and beauty of wine-producing regions while also enjoying the wines themselves. Running a successful wine tourism business requires a combination of factors, including a deep knowledge of the local wine industry, an understanding of hospitality and customer service, a passion for wine, and a commitment to sustainable practices.

A knowledge of the local wine industry is crucial to running a successful wine tourism business. This includes an understanding of the different types of wines produced in the region, the history and culture of winemaking in the area, and the various vineyards and wineries that visitors may want to visit. This knowledge can be gained through research, training, and experience in the wine industry.

An understanding of hospitality and customer service is essential for running a successful wine tourism business. This includes a focus on providing an exceptional customer experience, from the moment visitors arrive to the moment they leave. This may involve providing guided tours of local vineyards and wineries, offering wine tastings and food pairings, and providing transportation and accommodations for visitors.

A passion for wine is also important for running a successful wine tourism business. This passion can be shared with visitors through wine education, tasting notes, and recommendations for local wines to try.

A deep appreciation for the nuances of different wines and an ability to communicate that passion to others is key to creating a memorable wine tourism experience.

Most notably, a commitment to sustainable practices is becoming increasingly important in the wine tourism industry. This may include practices such as using organic or biodynamic farming methods, reducing waste and energy use, and supporting local communities and businesses. Sustainable practices not only help to preserve the natural environment and support local economies, but they can also appeal to consumers who are increasingly interested in ethical and sustainable practices.

Attaining Environmental Sustainability in Winemaking

There are multiple sustainable practices that businesses can adopt to reduce their impact on the environment and improve their overall sustainability. Simone Nulli Rinalducci, Director Ex. McLaren-Ex. Ferrari, Founder of Sustainability Success eloquently describes his career focus as "working efficiently and avoiding the waste of precious human, environmental, and economic resources" (Nulli Rinalducci, 2020). The premise of his approach to best practices includes three pillars including saving money, reducing waste, and conserving natural resources (Nulli Rinalducci, 2020).

Applying the McLaren-Ferrari ideology, a winery can invest in energy efficiency by implementing energy-efficient technologies, including LED lighting and HVAC systems, and reducing energy consumption through practices such as turning off lights and equipment when not in use. Implementing waste reduction programs, such as composting, recycling, and repurposing, to minimize the amount of waste generated by the business. By integrating water-saving technologies, such as low-flow toilets and faucets, and reducing water consumption through practices such as fixing leaks, the impact of water usage within the business is reduced. Another relative practice includes purchasing products and services from suppliers that have environmentally sustainable practices, such as using recycled materials and reducing waste. Wine producers are also exploring more sustainable packaging options such as lightweight glass bottles, reusable or recyclable containers, and alternative packaging materials like paper, aluminum, or bioplastics. Moreover, co-branding is a consideration the marketing department can develop. By encouraging employees to use alternative forms of transportation, such as cycling, public transit, or carpooling, and reducing the number of single-occupancy vehicle trips, will reduce employee carbon footprint. Measuring and reducing the business's carbon footprint through initiatives such as using renewable energy and reducing emissions from transportation and business operations also contributes to the overall impact the business has on the environment.

Renewable energy sources, such as solar or wind power, reduce the business's reliance on non-renewable energy sources and dependency on the electric grid. According to Wine Growers of Canada (2020), to become environmentally sustainable, a winery can implement several practices. Included are multiple considerations.

- Sustainable grape-growing practices include reducing the use of pesticides and fertilizers, conserving water, and using cover crops to maintain soil health.
- Energy-efficient operations such as using renewable energy sources incorporates solar or wind power and reducing energy consumption using energy-efficient lighting, heating, and cooling systems.
- Recycling and waste reduction involves recycling materials such as glass, paper, and plastic, and reducing waste by using biodegradable products and composting organic materials.
- Carbon footprint reduction considers shrinking greenhouse gas emissions through the use of energy-efficient equipment and transportation methods, and offsetting emissions through carbon credits or other initiatives.
- Water conservation encompasses reducing water usage in the production process and protecting water resources using sustainable irrigation practices.

By implementing these practices, a winery can become environmentally sustainable, reduce its impact on the environment, and improve its reputation as a responsible business.

Implications of Vineyard Parcels

The complexity of wine and the nuances in its taste have been a source of fascination for consumers and professional tasters. One of the most important factors in the quality and uniqueness of a wine is its origin, and winemakers take great care to guarantee the origin of their wine down to vineyard parcels that may only be a hundred meters apart. To better understand complexity in wine, the methods, and techniques that winemakers use to ensure the authenticity of their wines should be reviewed.

The first step in guaranteeing the origin of a wine is to establish a system of classification. This system can vary depending on the country or region in which the wine is produced, but it generally involves a set of rules and regulations that define the different types of wine and the conditions under which they can be produced. In some cases, the classification may be based on the grape variety used, while in other cases, it may be based on the geographic location of the vineyard. This is best exemplified in the European Wine Classification system.

The history of protecting a wine's origin dates back to the Bordeaux 1855 Classification. The Bordeaux 1855 Classification is a historic system of classification for red wines from the Bordeaux region of France. The classification was created for the Universal Exhibition held in Paris in 1855, which was intended to showcase the best wines of France to the world. The Bordeaux 1855 Classification was created by the Bordeaux Chamber of Commerce, and it has since become one of the most famous and influential wine classifications in the world.

The classification divided the top Bordeaux wines into five tiers, or "growths," based on their perceived quality and reputation at the time.

The top tier, known as the "First Growth," included five wines: Château Lafite Rothschild, Château Latour, Château Margaux, Château Haut-Brion, and Château Mouton Rothschild (which was elevated to First Growth status in 1973). The second tier, or "Second Growth," included 14 wines, while the third, fourth, and fifth tiers, or "Third Growth," "Fourth Growth," and "Fifth Growth," respectively, included a larger number of wines, (Stevenson, 2005).

The classification was based on the prices that the wines were fetching in the market at the time, with the most expensive wines being designated as First Growths, and so on. The classification was intended to be a one-time event, but it has endured to this day, with only one change in its history (the promotion of Château Mouton Rothschild to First Growth status).

The Bordeaux 1855 Classification has had a lasting impact on the Bordeaux wine industry and has helped to establish the reputations of many of the top Bordeaux châteaux, (Stevenson, 2005).

The system of classification has also been emulated by other wine regions around the world, demonstrating the enduring influence of Bordeaux on the world of wine. While the classification system may not be perfect or fully representative of the current quality of the wines, it remains an important part of Bordeaux's wine history and heritage.

Once the classification system is established, winemakers can begin to track the origin of their wines. This is done through a combination of physical and administrative measures. For example, each vineyard parcel may be assigned a unique identification number, and this number is used to track the grapes as they are harvested and processed. Administrative records are also kept documenting the various stages of the winemaking process, from the initial planting of the vines to the final bottling of the wine.

Another important factor in guaranteeing the origin of a wine is the use of geographical indicators. These indicators can include the name of the vineyard, the region in which it is located, or even the specific plot of land where the grapes were grown. These indicators are often included on the label of the wine, and they serve as a way for consumers to verify the authenticity of the wine.

To ensure that the wine is truly from the designated origin, winemakers may also implement strict quality control measures. For example, they may conduct regular inspections of the vineyards to ensure that the grapes are being grown under the proper conditions. They may also use advanced technology, such as satellite imagery, to monitor the growth of the vines and the ripening of the grapes.

Finally, winemakers may also rely on the expertise of local winemaking associations or regulatory bodies to ensure the authenticity of their wines.

These organizations may have strict guidelines and standards that winemakers must follow in order to use certain geographic indicators or labels on their wines. Winemakers go to great lengths to guarantee the origin of their wines down to vineyard parcels that may only be a hundred meters apart.

Through a combination of classification systems, physical and administrative measures, geographical indicators, quality control measures, and local expertise, winemakers can ensure the authenticity of their wines and provide consumers with a unique and enjoyable drinking experience. Furthermore, by achieving a better understanding of grape varieties best suited for the particular viticultural area will support a grape that is able to respond to the biodiversity of plant life, environment, and climate in a manner that requires less intervention.

Technology in an Eco-friendly Environment

Technology can assist in maintaining an eco-friendly environment in the vineyard in several ways. For example, precision farming, also known as precision agriculture, is a method included in managing vineyards that included technology such as drones, GPS, and sensors can be used to monitor and analyze the vineyard's soil and weather conditions. This information can be used to precisely apply water, fertilizer, and pesticides only where they are needed, reducing waste, and minimizing the impact on the environment.

Precision farming, also known as precision agriculture, is an approach to crop management that uses technology to optimize agricultural practices and increase productivity while minimizing waste and environmental impact.

This data is then analyzed to provide insights and inform decisions about when and where to apply inputs such as fertilizers, pesticides, and water, as well as when to harvest crops. The goal of precision farming is to maximize yields while minimizing inputs, reducing waste, and improving sustainability. By using precision farming techniques, farmers can improve their profitability, increase their efficiency, and reduce their environmental impact.

Many wineries are integrating alternative energy sources by adopting renewable energy sources such as solar, wind, and geothermal to power their operations. This reduces their carbon footprint and dependence on non-renewable energy sources. Furthermore, wineries are using technology to reduce water consumption by recycling and reusing water, using drip irrigation systems, and employing water-saving technologies.

Some wineries are using organic and biodynamic farming methods that avoid the use of synthetic fertilizers and pesticides. Organic farming is a method of farming that avoids the use of synthetic fertilizers, pesticides, and genetically modified organisms (GMOs). Instead, organic farmers rely on natural methods to maintain the health of the soil and crops, such as crop rotation, composting, and using natural pest control methods. Organic farming also emphasizes the use of renewable resources and conservation of soil and water.

Biodynamic farming is a method of farming that goes beyond organic farming by considering the broader ecological and spiritual aspects of farming. Biodynamic farming considers the farm as a holistic organism and aims to promote its health and vitality by using preparations made from plants, animals, and minerals. Biodynamic farmers also use planting and harvesting schedules based on lunar and astrological cycles.

Both organic and biodynamic farming methods prioritize the health of the soil, the environment, and the well-being of the animals and people involved in farming. While the two methods share many similarities, biodynamic farming incorporates more holistic and spiritual principles. Each method puts focus on building healthy soil and biodiversity in the vineyard, resulting in healthier vines and better-quality grapes.

Wineries are using technology to reduce waste by composting grape pomace and other organic material, and by recycling or repurposing packaging and other materials. By implementing these and other eco-friendly practices, wineries can reduce their environmental impact while producing high-quality wines. Technology plays a key role in helping wineries achieve these goals by providing the tools and data needed to make informed decisions about sustainable vineyard management.

Consumerism and Environmentalism's Effect on Business Strategies

Consumerism refers to the social and economic ideology that emphasizes the acquisition and consumption of goods and services as a way of life. It is a cultural and economic system that encourages people to buy more and more goods and services, often beyond what they actually need, in order to satisfy their desires and improve their social status.

Consumerism emerged in the twentieth century as a result of the rise of mass production, mass media, and the growth of the middle class. The idea of consumerism is closely linked to the concept of materialism, which emphasizes the importance of material possessions in life. Wine may fall into this category.

While consumerism has contributed to the growth of the economy and the improvement of living standards for many people, it has also been criticized for its negative effects on the environment, social inequality, and personal well-being. Some argue that consumerism promotes a throwaway culture that leads to waste, environmental degradation, and resource depletion. Others suggest that it fuels a sense of competition and status-seeking that can be detrimental to personal happiness and social harmony. Wine tourism is a link to better understand the business commitment to viticulture and viniculture.

Environmentalism has had a significant impact on the business of wine strategies, as sustainability and environmental responsibility have become increasingly important to consumers. Wineries are recognizing that being environmentally responsible not only benefits the planet but also provides a competitive advantage in the market.

One of the main ways that environmentalism has influenced wineries is through sustainable farming practices. This involves reducing the use of chemical fertilizers and pesticides, conserving water, and soil, and promoting biodiversity. Many wineries have switched to organic or biodynamic farming practices, which use natural methods to improve the health of the soil and vines, resulting in higher-quality grapes and wines.

In addition to sustainable farming, wineries are also implementing environmentally friendly production processes. Some wineries are even adopting circular economy models, where waste materials are reused or recycled, and the environmental impact of the production process is minimized.

Moreover, environmentalism has also influenced wineries' packaging and transportation strategies. Wineries are using eco-friendly packaging materials, such as recycled glass, and reducing the weight of their bottles to minimize the carbon footprint of transportation. Some wineries are also utilizing electric or hybrid vehicles for transportation, further reducing their carbon footprint.

Overall, the influence of environmentalism on winery business strategies has been significant. By adopting sustainable and environmentally responsible practices, wineries can differentiate themselves in a crowded market, attract environmentally conscious consumers, and help to protect the planet for future generations. Through tourism, visitors and consumers can witness positive change firsthand.

Community-based Environmental Business Strategies

The wine industry is a significant contributor to the global economy, with millions of people employed directly or indirectly. However, wine production can have significant environmental impacts, including land use change, water pollution, and greenhouse gas emissions. As such, community-based environmental business strategies can help mitigate these negative impacts and create a more sustainable wine industry.

One of the most effective ways to reduce the environmental impact of wine production is to adopt sustainable vineyard management practices. These practices can include reducing pesticide and herbicide use, using cover crops, and implementing water-saving techniques such as drip irrigation. Community-based initiatives can encourage wineries to adopt these practices by providing education and resources. By sourcing grapes locally can reduce the carbon footprint of wine production by minimizing transportation emissions. Community-based initiatives can encourage wineries to source grapes from nearby vineyards by creating networks of local growers and providing incentives for local sourcing.

Wineries consume a significant amount of energy, much of which comes from non-renewable sources. Community-based initiatives can encourage wineries to adopt renewable energy sources such as solar or wind power. These initiatives can provide education on renewable energy options, connect wineries with local renewable energy providers, and offer incentives for adopting renewable energy sources. Wine production generates a significant amount of waste, including grape skins, stems, and seeds. Community-based initiatives can encourage wineries to reduce waste by implementing composting programs and using waste as a source of energy through anaerobic digestion.

Finally, wine tourism is a significant contributor to the wine industry's economic growth. Community-based initiatives can encourage wineries to adopt ecotourism practices that minimize the impact on the environment. For example, wineries can provide electric vehicle charging stations for visitors and use environmentally friendly materials in their tasting rooms.

Community-based environmental business strategies can play a crucial role in creating a more sustainable wine industry. These strategies can help reduce the environmental impact of wine production while also providing economic benefits to local communities. By encouraging wineries to adopt sustainable practices, source grapes locally, adopt renewable energy, reduce waste, and embrace ecotourism, community-based initiatives can help create a more sustainable and prosperous wine industry.

Conclusion

Wine tourism has become a significant and growing segment of the travel industry, with many wineries around the world recognizing the value of offering visitors a unique and immersive experience that combines wine tasting with stunning natural settings and cultural experiences. The business of wine has also evolved, with winemakers focusing on sustainable practices to reduce their environmental impact, while also improving the quality of their products and maintaining the long-term viability of their vineyards.

Sustainability is becoming increasingly important to the wine industry, as climate change and other environmental challenges threaten the production of high-quality grapes. Many winemakers are implementing innovative solutions, such as organic and biodynamic farming, to reduce their carbon footprint and conserve natural resources, while also enhancing the biodiversity of their vineyards.

The impact of the environment and sustainability on the wine industry cannot be overstated, as these factors are critical to ensuring the longevity and success of winemaking operations. By adopting sustainable practices and promoting environmentally friendly tourism, wineries can not only protect the natural beauty and cultural heritage of their regions but also contribute to the economic development of their communities.

Overall, wine tourism, the business of wine, and sustainability are all interconnected and vital to the success of the wine industry. By working together to promote responsible tourism and sustainable winemaking practices, wineries can help to preserve the environment, support local communities, and offer visitors an unforgettable and authentic wine tasting experience.

References

Byrd, E. T., Canziani, B., Hsieh, Y., Debbage, K., & Sonmez, S. (2016). *Wine tourism: Motivating Visitors*. https://doi.org/10.3390/su141911831. Wine Tourism. (n.d.). Wine Tourism | Encyclopedia MDPI. https://encyclopedia.pub/entry/21769Szolnoki

Global status report on alcohol and health 2018 ISBN 978-92-4-156563-9 © World Health Organization 2018 Some rights reserved. This work is available under the Creative Commons Attribution-Non-commercial-ShareAlike 3.0 IGO licence (CC BY-NCSA 3.0 IGO; https://creativecommons.org/licenses/by-nc-sa/3.0/igo).

Gomez, M., Lopez, C., & Molina, A. (2015). A model of tourism destination brand equity: The case of wine tourism destinations in Spain. *Tourism Management, 51*, 210–222.

López, A. M. (2023). *Market size of wine tourism worldwide in 2020, with forecast for 2030.* Statista. https://www.statista.com/statistics/912835/market-size-enotourism-worldwide/

Maykish, A., Rex, R., & Sikalidis, A. K. (2021, January 8). Organic winemaking and its subsets; biodynamic, natural, and clean wine in California. *Foods, 10*(1), 127. https://doi.org/10.3390/foods10010127.PMID: 33430151;PMCID:PMC7827467

Nulli Rinalducci, S. (2020). *Sustainability success.* Sustainable Business Practices. https://sustainability-success.com/sustainable-business-practices/

State of the world Vine and Wine Sector 2021; International Organization of Vine and Wine Intergovernmental Organization (OIV). (2022, April). Retrieved March 2023. https://www.oiv.int/sites/default/files/documents/eng-state-of-the-world-vine-and-wine-sector-april-2022-v6_0.pdf

Statista; Travel and Tourism, Total Economic Contribution Worldwide; Retrieved March 2023. https://www.statista.com/statistics/233223/travel-and-tourism--total-economic-contribution-worldwide/

Stevenson, T. (2005). *The Sotheby's wine encyclopedia* (4th ed.). Dorling Kindersley.

United Nations. (2023). *Sustainable development goals.* United Nations. https://www.un.org/en/academic-impact/sustainability

University of Mary Washington. (2023). *Office of sustainability.* University of Mary Washington. https://sustainability.umw.edu/areas-of-sustainability/economic-sustainability

UNWTO. (2023). *Innovation, sustainability and creativity: UNWTO global wine tourism conference 2022.* UNWTO. https://www.unwto.org/news/unwto-wine-tourism-conference-innovation-sustainability-creativity

UNWTO. (2023). *Academic impact sustainability.* UNWTO. https://www.un.org/en/academic-impact/sustainability

UNWTO. (2023). *Gastronomy and wine tourism.* UNWTO. https://www.unwto.org/gastronomy-wine-tourism#:~:text=Whilst%20Eno%2Dtourism%20(Wine%20Tourism,at%20or%20near%20the%20source

UNWTO. (2023). *Wine tourism—A growing tourism segment.* UNWTO. https://www.unwto.org/archive/global/press-release/2016-09-09/wine-tourism-growing-tourism-segment

Wilson, A. (2021, April). Retrieved January 2023. https://eco.ca/blog/what-is-sustainability-and-why-is-it-important/

Wine Growers of Canada. (2020). *Wine growers of Canada.* Environmental Sustainability Principles. https://www.winegrowerscanada.ca/member-info/sustainable-practices/

Wine Producing Countries; Retrieved March 2023. https://worldpopulationreview.com/country-rankings/wine-producing-countries
WineTourism.com. (n.d.). *What is wine tourism—Enotourism—Vinitourism—Gastrotourism?* WineTourism.com. https://www.winetourism.com/synonyms-wine-tourism-enotourism-vinitourism/. Admin published November 20, 2011.
World's Best Wines. (2023). *Wines from all around the world*. World's Best Wines. https://worldsbestwines.eu/
World Commission on Environment and Development . (1987). *1987: Brundtland report*. World Commission on Environment and Development. https://www.are.admin.ch/are/en/home/media/publications/sustainable-development/brundtland-report.html

Discovering *Enocotourism* in the Ribera del Duero Appellation: A New Concept Combining Wine Tourism and Ecotourism for a Unique Experience

Rosana Fuentes-Fernández,
María del Carmen González-Velasco,
and Marcos González-Fernández

The Ribera Del Duero's Environmental Dimension

Organic wines have gained popularity over the past few years due to consumers' growing awareness of their impact on the environment and personal health. These wines are made from grapes that are cultivated without the use of synthetic pesticides, herbicides, or fertilizers. The winemaking process also avoids the use of artificial additives and preservatives. Instead, natural and sustainable farming and winemaking

R. Fuentes-Fernández (✉) · M. del Carmen González-Velasco ·
M. González-Fernández
Department of Business Management and Economics, University of León, León, Spain
e-mail: rosana.fuentes@unileon.es

M. del Carmen González-Velasco
e-mail: mcgonv@unileon.es

M. González-Fernández
e-mail: mgonf@unileon.es

J. Martínez-Falcó et al. (eds.), *Wine Tourism and Sustainability*,
https://doi.org/10.1007/978-3-031-48937-2_7

practices are used, which benefits both the environment and the wine's quality.

Organic farming practices foster enhanced biodiversity and promote healthier soil quality when compared to conventional vineyards. This healthier ecosystem benefits the grapevines, producing high-quality grapes that are more resistant to pests and diseases. In turn, this translates to better-tasting wines that truly reflect the unique terroir of the vineyard. Organic wines have also been found to contain higher levels of antioxidants and polyphenols, which are beneficial for human health.

The production of organic wines is a more sustainable approach to winemaking, as it reduces the use of synthetic chemicals and promotes natural and environmentally friendly farming practices. This approach also supports the local economy by creating jobs and promoting rural development. As a result, organic wines have become increasingly popular among consumers who are looking for healthier and more sustainable options.

However organic wines not only offer numerous benefits in terms of taste and health, but also align with the principles of ecotourism. According to the World Tourism Organization (2002), ecotourism is a form of sustainable tourism that focuses on responsible travel to natural areas and aims to minimize the negative impacts on the environment while promoting conservation efforts and providing educational and cultural experiences to visitors. The adoption of organic viticulture practices by vineyards and wineries can contribute towards preserving natural resources, reduce the use of harmful chemicals, and promote biodiversity, all of which align with the principles of ecotourism. Depending on the landscape and local habitat conditions, organic viticulture can have a positive impact on carabid beetles. The authors suggest that organic viticulture can contribute to the preservation of biodiversity, which aligns with the principles of ecotourism as mentioned before. Thus, the growing demand for organic wines and the adoption of sustainable practices by vineyards and wineries can provide unique ecotourism experiences to visitors while preserving both the environment and cultural heritage.

Regarding the environmental dimension in the wine sector, it refers to the impact of vineyard and winery activities on the environment, as well as to the adoption of practices that contribute to the preservation and protection of natural resources. This dimension covers issues such as soil management, water conservation, energy efficiency, waste reduction, and the use of environmentally friendly materials, among others. According to several authors, there is a growing trend towards adopting environmentally sustainable practices is becoming increasingly important for wineries, both to respond to consumer demand and to comply with regulations and certifications. For example, when firms adopt sustainability practices, "they become more valued in the market" (Lourenço et al., 2014) and "see opportunities to achieve competitive advantage" (Babu et al., 2018; Gurtu et al., 2017).

In this context, we focus in a specific wine region, the Ribera del Duero, located in the north of Spain. The region that runs parallel to the Duero River is famous for producing top-notch red wines, predominantly from the Tempranillo grape. The region's unique terroir, which includes a combination of altitude, climate, and soil composition, contributes to the distinct character of the wines produced there. In recent years, the Ribera del Duero wine region has made significant efforts to promote sustainable and environmentally friendly practices. For example, many vineyards have adopted organic and biodynamic viticulture practices, which use natural and sustainable farming techniques to promote soil health, biodiversity, and grape quality. The region's wineries also prioritize energy efficiency and waste reduction in their winemaking processes. These efforts have not only resulted in the production of high-quality wines, renowned for their unique character shaped by the region's altitude, climate, and soil composition, predominantly from the Tempranillo grape. Additionally, the commitment to sustainable practices, such as organic and biodynamic viticulture, has not only elevated grape quality but also positively impacted the environment. Many vineyards in the Ribera del Duero region prioritize natural and sustainable farming techniques, promoting soil health and biodiversity. Furthermore, the region's wineries actively focus on energy efficiency and waste reduction in their winemaking processes,demonstrating a holistic approach to environmentally friendly practices. These initiatives not only contribute to the excellence of the wines but also foster a positive impact on the surrounding ecosystem and the well-being of the local community.

The Ribera del Duero wine region is committed to preserving its natural resources, as well as promoting its cultural heritage. Visitors to the Ribera del Duero wine region can participate in wine tours, tastings, and cultural experiences that showcase the region's unique character and highlight its commitment to sustainable tourism practices.

Combining Enotourism and Ecotourism: The Rise of Enocotourism

Enotourism, also known as wine tourism, is a growing trend in the tourism industry that offers visitors the opportunity to experience the culture and traditions of wine production in different regions around the world. As stated by Molina et al. (2015), the industry of wine tourism is expanding and becoming an important factor in the growth of rural areas. The success of this industry relies on various factors such as viticulture, the availability of wine tourism products (e.g., wineries, wine routes, and festivals), the connection to regional development, and the significance of tourist attractions. Thus, organic wines produced through sustainable farming practices that promote biodiversity and healthier soil, are a perfect fit for enotourism, as they reflect the values of quality, authenticity, and sustainability that are at the core of this form of tourism. According to Magnusson et al. (2003), the production of organic wine is closely linked to an awareness of sustainability.

Ecotourism and enotourism are two distinct but related forms of sustainable tourism that share similar goals and principles. Based on the information provided by the International Ecotourism Society: "Ecotourism aims to promote responsible travel to natural areas that conserves the environment and sustains the well-being of local people" (2015), while enotourism focuses on promoting responsible travel to wine regions that support the local wine industry and culture (Sangpikul, 2010). Enotourism is not only about tasting wine, but also about experiencing the wine region's unique cultural, historical, and social features, as noted by Tommasetti and Festa (2014) wine tourism has garnered significant attention and is now widely acknowledged as one of the most fascinating aspects of Italy's tourism industry. Enotourism offers visitors the opportunity to learn about the wine region's natural environment, biodiversity, and sustainable wine production practices, as well as to support local communities and economies through wine-related activities. Furthermore, the level of trust that consumers have in a company or product plays a crucial role in their willingness to support environmentally friendly options. This has been highlighted in various studies, including those conducted by Bonn et al. (2016), Chen and Chang (2013), and Ponnapureddy et al. (2017). Overall, enotourism represents a sustainable and responsible form of tourism that supports the conservation of the natural and cultural legacy of wine regions while promoting economic and social development.

Hence, what is the term for the connection between eco-friendly wine tourism and ecotourism? Additionally, how does this connection relate to conserving the natural and cultural resources found in wine-producing areas? The term "enocotourism" is the answer to these questions (Fuentes-Fernández et al., 2022). The concept merges the ideas of wine tourism and sustainable tourism, highlighting the significance of safeguarding both natural and cultural resources of the winegrowing regions while promoting local economies and businesses. By creating this term, we can more clearly define and promote the idea of environmentally friendly wine tourism, which benefits both the visitors and the communities they visit. Enocotourism, therefore, represents a holistic and inclusive approach to wine tourism that values sustainability and quality tourism experiences.

Organic wines are a perfect fit for enocotourism as they are produced through sustainable farming practices that promote biodiversity, healthier soil, and higher-quality grapes. The Ribera del Duero wine region in Spain is an excellent example of a destination that has embraced wine tourism (enotourism) and sustainable practices (ecotourism), creating a unique and attractive destination for wine enthusiasts and eco-conscious travelers. With its beautiful landscapes, rich cultural heritage, and world-class wineries, enotourism has gained immense popularity in the Ribera del Duero region, drawing tourists from every corner of the globe, but also it is a prime example of an area that has embraced sustainable wine practices. Many of the region's vineyards have adopted organic and biodynamic farming practices, which prioritize the use of natural resources and environmentally friendly techniques.

This new approach emphasizes the importance of environmentally friendly agricultural practices and quality, local tourism. In addition to providing knowledge about wine, it is also important to educate wine tourists about the importance of preserving the environment and the cultural heritage of the wine region. In Ribera del Duero, sustainable wine tourism practices are crucial for the preservation of the vineyard's unique natural environment and the region's cultural heritage. As stated by Ko and Liu (2016), a pleasant setting can aid in comprehending customers' perspectives on wineries' environmental initiatives and facilitate the promotion of their eco-friendly business through targeted marketing strategies.

Furthermore, sustainable wine tourism practices in Ribera del Duero can also have a positive economic impact on the region. According to a study by Fuentes-Fernández and colleagues in 2022, ecological farming has the potential to moderate the correlation between wine tourism and economic success in Ribera del Duero. This finding highlights the potential for sustainable wine tourism practices to support the region's economic development while preserving its natural and cultural heritage.

In conclusion, enocotourism practices promote a sustainable approach to wine tourism that values quality experiences, environmental protection, and cultural heritage preservation. In Ribera del Duero, sustainable wine tourism practices are essential for preserving the vineyard's unique natural environment, protecting ecosystem services, and fostering economic development while preserving the region's cultural heritage.

Data and Methodology

The study focused on wineries Situated in Spain's the Ribera del Duero region, this area encompasses 19 municipalities in Soria, 60 in Burgos, 4 in Segovia, and 20 in Valladolid. There were 239 active businesses that were engaged in the production of wine and were categorized under the National Classification of Economic Activities code 1102 (Fuentes-Fernández, et al., 2022).

The empirical analysis utilized two primary variables: wine tourism (WT) and ecological wines (EW). These variables were considered dichotomous, with a value of 1 indicating that a winery engaged in either wine tourism or ecological activities, and 0 indicating that they did not. The information gathered from the wineries' corporate websites was used to determine whether or not they engaged in these activities.

In order to establish if there is a significant correlation between "organic wine" and "wine tourism," a chi-square ($\chi 2$) test was utilized to conduct a correlation analysis. This is a statistical measure of independence between two categorical variables. The reason for using the chi-square test is that both variables are binary, meaning they can take on only two values (yes or no). A test of this kind is frequently employed to evaluate the correlation between two groups of categorical data, especially when the information is non-parametric and does not fulfill the prerequisites for parametric tests like the t-test or ANOVA.

The process of conducting a chi-square test involves computing the expected frequencies for each category in the contingency Table 1. This computation is based on the assumption that the two variables are independent of each other. The expected frequencies are then compared to the observed frequencies, and the difference between them is used to calculate the value of the chi-square statistic.

This statistic is then compared to a critical value from a chi-square distribution, which has degrees of freedom equal to (r-1) (c-1), where r is the number of rows and c is the number of columns in the contingency Table 1. If the calculated chi-square statistic is greater than the critical value, then it can be concluded that the two variables have a significant relationship and the null hypothesis of independence must be rejected. However, if the calculated chi-square statistic is less than the critical value, then it can be concluded that there is no significant relationship between the two variables and the null hypothesis of independence cannot be rejected. The significance of this type of test lies in its ability to determine the relationship between two variables without making any assumptions about the underlying distribution of the data.

This method is especially helpful when dealing with data that is not normally distributed or when the sample size is limited. It also has the ability to reveal information about the correlation between variables, such as whether it's positive or negative, and the degree of the correlation.

Results

The chi-square value obtained from Pearson is 0.130 with 1 degree of freedom, and the bilateral asymptotic significance is 0.718. Subtracting the asymptotic significance from 1 gives a significance level of 0.282, which indicates that there is a 28.2% chance of a relationship between two variables occurring by chance.

The confidence level is 71.8%, meaning that there is a 71.8% probability that the relationship between the two variables is not due to chance. Since the significance value is greater than the level of significance (0.718 > 0.05), the null hypothesis cannot be rejected. Therefore, there is no statistical proof of association between the variables under analysis.

However, it was found that 49 wineries produce both organic wines and offer wine tourism services, suggesting a growing interest in offering

Table 1 Chi-square tests

	Value	Df	Asymptotic significance (two-tailed)	Exact significance (two-tailed)	Exact significance (one-tailed)
Pearson's chi-squared	0.130[a]	1	0.718		
Continuity correction[b]	0.053	1	0.817		
Likelihood ratio	0.130	1	0.718		
Fisher's exact test				0.795	0.409
Linear-by-linear association	0.130	1	0.719		
Number of valid cases	239				

a. 0 cells (0.0%) had an expected count less than 5. The minimum expected count is 53.39

Source SPSS

tourism experiences related to wine and environmental sustainability. This may be a sign that consumers are increasingly demanding organic wines and wine-related tourism experiences, suggesting that visiting wineries has become a significant pastime in the Ribera del Duero region. Overall, wineries are responding to this demand by adapting their commercial and marketing strategies.

Conclusions

This book chapter delves into the interplay between the environmental and tourism dimensions of the Ribera del Duero's wine region in Spain. It specifically explores the convergence of enotourism and ecotourism, leading to the proposal of a novel term: enocotourism. By combining these two concepts, this chapter sheds light on the unique sustainability and cultural aspects that arise when wine tourism and environmental preservation intersect in this renowned region.

In the wine region of the Ribera del Duero, there is a strong commitment to preserving the natural resources and cultural heritage of the area. The region also strives to promote sustainable development and provide visitors with unique experiences that showcase the character of the region and its dedication to sustainable tourism practices. Wine tourism is an important and growing segment of the tourism industry, and it plays a significant role in driving regional economic development. The text also explores the combination of ecotourism and enotourism, highlighting the compatibility of organic wines with enotourism due to their emphasis on quality, authenticity, and sustainability. Enotourism has become increasingly popular in Ribera del Duero, drawing visitors from around the world who are interested in eco-friendly practices and the production of wine.

Additionally, wine tourism is a multifaceted experience that involves not only wine tasting, but also cultural, culinary, and environmental elements. The combination of enotourism and ecotourism in Ribera del Duero offers visitors a unique and holistic experience that goes beyond just wine tasting, incorporating the region's cultural and natural heritage, sustainable development practices, and local cuisine. This approach not only enhances the visitor experience but also contributes to the regional economic development of the area.

In conclusion, the Ribera del Duero wine region in Spain is a good example of a sustainable and eco-friendly destination for wine tourism. However, there is still much to be done to make it a more visible and representative example. By implementing organic and biodynamic viticulture practices, the region is promoting soil health, biodiversity, and grape quality while also reducing the use of synthetic chemicals. This commitment to sustainability is reflected in the higher levels of antioxidants and polyphenols found in the organic wines produced in the region, which are not only beneficial for human health but also contribute to the unique character and authenticity of the wines. The International Ecotourism Society outlines a set of principles for ecotourism, which include reducing negative impact, promoting environmental and cultural awareness and respect, ensuring positive experiences for visitors and hosts, contributing directly to conservation efforts, and providing financial benefits and empowerment for local communities.

Wineries in the region should also consider incorporating educational components into their wine tourism offerings, as wine knowledge and wine region visits can significantly impact consumers' demand for local wines. Moreover, following best practices in wine tourism, such as focusing on the quality of the visitor experience, developing partnerships and collaborations with other local businesses and organizations, and integrating sustainability and environmental stewardship into tourism planning and management, can further enhance the appeal of the Ribera del Duero wine region for wine enthusiasts from around the world. Tommasetti and Festa (2014) highlight the importance of effective communication and collaboration between ecotourism stakeholders, including government agencies, local communities, and tourists, to ensure that sustainable practices are implemented and enforced. Overall, the combination of enotourism and ecotourism in this unique destination offers a valuable and sustainable model for the wine tourism industry.

The bivariate analysis conducted on the two variables did not reveal any statistically significant association between them, indicating that they may be independent. Despite statistical results, Ribera del Duero remains an excellent destination to conduct such an analysis due to the aforementioned factors in terms of organic and biodynamic viticulture practices. Furthermore, the region's global prominence as a reference point in the wine industry solidifies its significance as a prime location for examining the intersection of sustainability and tourism.

However, the observation that almost 50 wineries produce organic wines and offer wine tourism services indicates a trend towards incorporating sustainable and eco-friendly practices in the wine industry. The discovery implies that wine tourism plays a significant role in the economy of the Ribera del Duero area, and the vineyards are adjusting to fulfill the increasing need for eco-friendly and sustainable wine merchandise and experiences. Therefore, it may be important for wineries to continue to invest in wine tourism and incorporate educational components to attract consumers interested in sustainable and eco-friendly practices in the wine industry.

Despite the relatively low number of wineries offering enocotourism services, it is worth noting that this represents a growing trend in the region. The significance of constantly monitoring and evaluating ecotourism operations to determine their effect on the environment and local communities, as well as to make necessary modifications to guarantee sustainable practices, is underlined in various studies, including Bonn et al. (2016), Chen and Chang (2013), and Ponnapureddy et al. (2017). This trend is likely a response to increasing demand from consumers who are interested in sustainable and eco-friendly tourism experiences. Moreover, the research carried out by Fuentes-Fernández et al. (2022) emphasizes the potential of sustainable wine tourism practices in Ribera del Duero to have a positive economic impact on the region. Their research illustrates that ecological agriculture can moderate the link between wine tourism and economic performance, underscoring the importance of sustainable practices in supporting the region's economic development while preserving its natural and cultural heritage. Overall, the combination of organic and biodynamic viticulture practices, enotourism, and ecotourism in the Ribera del Duero wine region offers a valuable and sustainable model for the wine tourism industry.

Wineries that incorporate these practices and focus on providing quality visitor experiences while integrating sustainability and environmental stewardship into their tourism planning and management are well-positioned to meet the growing demand for sustainable and eco-friendly wine products and experiences. As more wineries begin to offer enocotourism services, it is likely that the Ribera del Duero Appellation will become an even more popular destination for wine and tourism enthusiasts.

However, it is important to note that the development of enocotourism must be done in a sustainable manner. The wine industry must implement sustainable practices and minimize its environmental footprint in order to ensure that the region remains an attractive destination for future generations. By promoting enocotourism, the Ribera del Duero Appellation is moving in the right direction to ensure the long-term sustainability of the wine industry in the region.

References

Babu, D. E., Kaur, A., & Rajendran, C. (2018). Sustainability practices in tourism supply chain: Importance performance analysis. *Benchmarking, 25*(4), 1148–1170. https://doi.org/10.1108/BIJ-06-2016-0084

Bonn, M. A., Cronin Jr., J., & Cho, M. (2016). Do environmental sustainable practices of organic wine suppliers affect consumers' behavioral intentions? The moderating role of trust. *Cornell Hospitality Quarterly, 57*(1), 21–37.

Chen, Y-S., & Chang, C-H. (2013). Greenwash and green trust: The mediation effects of green consumer confusion and green perceived risk. *Journal of Business Ethics, 114*(3), 489–500.

Fuentes-Fernández, R., Martínez-Falcó, J., Sánchez-García, E., & Marco-Lajara, B. (2022). Does ecological agriculture moderate the relationship between wine tourism and economic performance? A structural equation analysis applied to the Ribera del Duero wine context. *Agriculture, 12*(12), 2143. https://doi.org/10.3390/agriculture12122143

Gurtu, A., Searcy, C., & Jaber, M. Y. (2017). Sustainable supply chains. In M. Khan, M. Hussain, & M. Ajmal (Eds.), *Green Supply Chain Management for Sustainable Business Practice* (pp. 1–26). IGI Global. https://doi.org/10.4018/978-1-5225-0635-5.ch001

International Ecotourism Society. (2015). *What is ecotourism?* Retrieved from https://ecotourism.org/what-is-ecotourism/

Ko, W. W., & Gordon, L. (2016). Environmental strategy and competitive advantage: The role of small- and medium-sized enterprises' dynamic capabilities. *Business Strategy and Environment, 26*(5), 584–596. https://doi.org/10.1002/bse.1938

Ko, W. W., & Liu, G. (2006). Environmental Strategy and Competitive Advantage: The Role of Small- and Medium- Sized enterprises' Dynamic Capabilities: Environmental Strategy and Competitive Advantage. *Business Strategy and the Environment, 26*(5). https://doi.org/10.1002/bse.1938

Lourenço, I. C., Callen, J. L., Branco, M. C., & Curto, J. D. (2014). The value relevance of reputation for sustainability leadership. *Journal of Business Ethics, 119*, 17–28. https://doi.org/10.1007/s10551-012-1617-7

Magnusson, M. K., Arvola, A., Koivisto, U.-K., Åberg, L., & Sjödén, P.-O. (2003). Choice of organic foods is related to perceived consequences for human health and to environmentally friendly behavior. *Appetite, 40*(2), 109–117.

Molina, A., Gómez, M., González-Díaz, B., & Esteban, A. (2015). Market segmentation in wine tourism: Strategies for wineries and destinations in Spain. *Journal of Wine Research, 26*(3), 192–224. https://doi.org/10.1080/09571264.2015.1051218

Ponnapureddy, S., Priskin, J., Ohnmacht, T., Vinzenz, F., & Wirth, W. (2017). The influence of trust perceptions on German tourists' intention to book a sustainable hotel: A new approach to analyzing marketing information. *Journal of Sustainable Tourism, 25*(70), 970–988.

Sangpikul, A. (2010). Marketing ecotourism though the internet a case of ecoturism business in Thailand. *International Journal of Hospitality and Tourism Administration, 11*(2), 107–137. ISSN 1525–6480.

Tommasetti, A., & Festa, G. (2014). An analysis of wine tourism in Italy from a strategic service-based perspective. *Service Science, 6*(2), 122–135. https://doi.org/10.1287/serv.2014.0070

World Tourism Organization. (2002). Global code of ethics for tourism. Retrieved from https://www.unwto.org/global-ethics-tourism

Social-Cultural Impact of Wine Tourism

Exploring the Intersection Between Geographical Indications and Sustainable Wine Tourism: The Case of Le Colline del Prosecco di Conegliano e Valdobbiadene

Francesca Checchinato, Cinzia Colapinto, Vladi Finotto, Christine Mauracher, and Chiara Rinaldi

Introduction

Successful wine regions are increasingly diversifying their economies by creating experiences around wine that can support sustainable tourism development (Santos et al., 2019). Literature on wine tourism has developed since the mid-90 s and generally focused on "old world" European wine-producing countries focusing more on wine as a product versus "new world" wine-producing countries, which tend to provide more varied wine tourism experiences (Carvalho et al., 2021). However, it

F. Checchinato · C. Colapinto · V. Finotto · C. Mauracher · C. Rinaldi (✉)
Agrifood Management and Innovation Lab, Department of Management, Ca' Foscari University of Venice, Venice, Italy
e-mail: chiara.rinaldi@unive.it

F. Checchinato
e-mail: checchin@unive.it

C. Colapinto
e-mail: cinzia.colapinto@unive.it; c.colapinto@ipag.fr

J. Martínez-Falcó et al. (eds.), *Wine Tourism and Sustainability*,
https://doi.org/10.1007/978-3-031-48937-2_8

appears that also "old world" countries are increasingly leveraging wine and related landscapes to attract cultural tourists. This is particularly true in wine regions that acquired the UNESCO World Heritage List (WHL) status as Cultural Landscapes, such as Champagne, Langhe-Roero and Monferrato, or Prosecco. Indeed, UNESCO recognition focuses on landscape, offering the opportunity to strengthen place-product associations for the interested territories. Therefore, this chapter addresses the intersection between UNESCO WHL and the Prosecco Geographical Indication (GI), by focusing on UNESCO's potential as leverage for wine tourism development.

Literature Review

According to Getz and Brown (2006, p.147) "wine tourism is, simultaneously, a form of consumer behaviour, a strategy by which destinations develop and market wine-related attractions and imagery, and a marketing opportunity for wineries to educate, and to sell their products, directly to consumers." Wine already represents a motivation to attract tourists to certain regions, but places inscripted to the UNESCO WHL have a further element strengthening the destination's profile due to UNESCO's prestigious brand name (Sigala, 2019).

Indeed, a "winescape"—a concept that summarizes natural and cultural elements of the wine regions typical landscape (Carvalho et al., 2021)—that further obtains UNESCO WHL recognition as a Cultural Landscape, can attract both "wine interested" tourists, as well as cultural tourists with diverse levels of interest and involvement in wine.

There are different perspectives concerning the UNESCO impact on tourism development. In an extensive literature review, Patuelli (2013)

V. Finotto
e-mail: vfinotto@unive.it

C. Mauracher
e-mail: maurache@unive.it

maintains that extant studies converge towards four outcomes: positive effects; positive but relatively limited effects; insignificant effects on tourism but relevant effects in terms of heritage protection; negative effects. Accordingly, it appears that results are strongly influenced by the contexts in which research was carried out. Moreover, literature signals differences in the effects of UNESCO recognition in terms of attracting international or domestic tourists (Yang et al., 2019), as international tourists might be more likely to select destinations relying on indicators such as UNESCO WHL.

Successful local products such as wine or relevant cultural tourism attractions such as the sites distinguished by the UNESCO recognition might be a crucial leverage to bring tourists to a place. However, having attractive resources is not enough for a place to become a tourist destination. Indeed, the existence of infrastructures aimed at incoming visitors proved to be pivotal for the deployment of effective strategies centered on tourist services and hospitality (Russo & van der Borg, 2002). Moving from this literature, this study focuses on the Valdobbiadene Conegliano Prosecco DOCG area to determine strengths and weaknesses of this territory in terms of its capacity to manage wine tourism development.

Background

In 2009, the Prosecco GI production area was expanded to include different provinces in Veneto and in the neighboring region of Friuli Venezia Giulia. The expansion created two types of Prosecco brands (DOCG and DOC). These changes contributed to make Prosecco a global commercial success with a steady growth in production which in 2022 exceeded 750 million bottles (Rinaldi et al., 2022). In 2019 the "Prosecco Hills of Conegliano and Valdobbiadene" site—encompassing the DOCG area—was inscribed in the World Heritage List as a Cultural Landscape linked to a distinctive winemaking tradition. In the area, the work of winegrowers contributed to creating a unique scenery. The site encompasses the area that ranges from the municipality of Valdobbiadene to the municipality of Vittorio Veneto descending towards Conegliano and includes a significant portion of the GI winegrowing landscape where the Conegliano Valdobbiadene Prosecco Superiore GI is produced.

The benefits of the UNESCO recognition on tourism development became immediately visible. Cultural tourism represents a growing trend, especially among World Heritage Sites. Until the COVID-19 pandemic in 2020, in fact, there was a steady increase of tourism in the area. In 2021, tourist arrivals grew + 66,3% compared to 2020, even though they are still -23,5% compared to 2019 (Distretto del Conegliano Valdobbiadene Prosecco DOCG, 2022). Despite the setback caused by the pandemic, interviewees agree that the trend of visitors in the last five years is very positive. In fact, 67% of wineries declare that they have recorded overall growth. Regarding the wine tourism offer, 89% of the 170 wineries in the Conegliano Valdobbiadene Prosecco DOCG District have a direct shop and 81% have a tasting room. The share of companies offering catering and accommodation services is respectively 7% and 12%.

Method

A qualitative research design underpinned our study of the case: the collection of qualitative evidence from a variety of primary and secondary sources was structured and interpreted through the methods and practices recommended by Gioia et al. (2013). The Gioia method is a qualitative research method that involves a rigorous and systematic approach to analyze data emerging from textual sources such as interviews, focus groups, and documents. This method is particularly useful to analyze complex and multidimensional data, and it is used to capture relevant and recurrent concepts over the interview in an inductive way. Data collection and subsequent analysis allowed us to develop an in-depth understanding of the dynamics underlying tourism management and hospitality approaches in the selected area before and after the UNESCO recognition.

Data collection relied eminently on interviews to representatives of eight wineries and two institutional actors in the area (to anonymize our data we refer to simple labels, such as winery 1 or institutional actor 1), as well as secondary sources represented by Consortia reports, websites, newspaper articles, and archival records. The interviewees occupied different positions in their organizations, ensuring a multiplicity of perspectives with regard to tourism development in the area, the opportunities it opened, as well as the actions to be deployed to consolidate flows and to make wine tourism a driver of economic and social development. We interviewed the director of the Consortium for the Protection of Prosecco of Conegliano Valdobbiadene and the site manager of the "Association for the Heritage of the Prosecco Hills of Conegliano and Valdobbiadene." As far as the firms are concerned, we interviewed informants acting as winemakers, public relations management, and hospitality managers of local wineries.

Each interview was recorded with the consent of each subject and later transcribed verbatim and analyzed following the steps of the well-known Gioia method.

Findings and Discussion

Prosecco wine global success had a visible impact on the rural landscape of the region, which was at risk of depopulation and abandonment in the first half of the twentieth century. Later, the area was revived thanks to the entrepreneurial spirit of local winemakers in the aftermath of the Second World War (Gobbato, 2009). The local wine, a white, sparkling, and accessible one, encountered rapidly the favor of national and international consumers, determining the gradual consolidation of dedicated agriculture and wine production. As a result, the area was transformed: vineyards climbing the irregular hills of the site are a testimony of how human practices and labor can interact with nature, creating sites that are environmentally unique and culturally rich.

The UNESCO recognition formalized the status of a distinctive site, laying the ground for the development of a thriving tourism sector, and triggering an increase in visitor numbers. The impact is not only quantitative though: the Hills seem to attract flows of qualified tourists when it comes to motivations, behavior, and interests towards the area.

Different Type of Tourists

It is widely acknowledged that wine tourism accounts for an important part of the tourist flows in Veneto. Our interviews pointed out that tourism in the Hills has evolved thanks to the growing international success of Prosecco, as well as the strong connection between the beauty of the landscape and the quality of this famous sparkling wine.

Indeed, an informant mentioned an increase of foreign tourists in the Hills: *"Many international tourists have visited our hills during the last years. In my opinion this is related to UNESCO recognition" (Winery 5).* Another one indicated that the tourists themselves appear different than usual ones that used to arrive in the area just to buy wine. *"What we see now is a different type of tourist that was not there before, and I think this has to do with UNESCO"* (Winery 4).

Another interviewee stressed the curiosity of the incoming visitors and their willingness to explore in depth the characteristics of the different sub-areas encompassing the Hills and the distinctive characteristics of local wineries: *"a curious tourist visiting the territory to acquire knowledge of the area and the differences among wineries and territories" (Winery 7).*

In the last few years tourists came from several countries, with an increase of flows originating from distant countries, while in the recent past international presences were eminently those from neighboring European states. Secondary data confirmed the growth of tourism in the area, as reported in the Background section. Tourists' expectations, in turn, appear to be quite sophisticated: experiences sought are not elementary in nature, such as consuming wine in cellars and learning about different wines and brands' characteristics. Rather, they converge towards asking for articulated accounts on, and visits of, production processes, in-depth views on the evolution of production cultures and local material cultures, and detailed reconstructions of the history of wineries, to grasp the social fabric underlying the evolution of the agro-industrial structure of the area.

Moreover, tourists prefer to visit and travel around, to enjoy the landscape. Finally, tourist demand is no longer only seasonal and concentrated in spring and summer: it extends the period of offering, with a positive impact for many restaurants and accommodation companies.

Whether the reason is only the UNESCO recognition or mainly depends on Prosecco wine is not easy to determine, as both drivers emerged from our interviews. Indeed, some interviewees believed that Prosecco was the main motivation tourism had increased:

> I don't think that tourism increase was due to UNESCO. Prosecco represents a strong attraction, and wine tourism itineraries started over fifteen years ago - contributing to a constant tourism growth. (Winery 2)

Others believe that UNESCO contributed to bringing a different type of tourism in the area, particularly in terms of international audience.

> We have experienced in the last years (2019 and then again post COVID19) tourists coming from different countries that we did not see before, such as Japanese, Indians, Israelis and so on. I think this might be an effect of UNESCO recognition. (Winery 8)

This supports Yang et al. (2019) findings: UNESCO recognition contributes to tourism development, enhancing the awareness of certain places in foreign countries.

UNESCO Recognition: A Collective Effort to embrace the Challenge

The UNESCO recognition influences the sense of belonging in a community by sharing values and norms and respecting the responsibility of every individual in contributing to the common good. Interviews highlighted that it "*enhances the territorial identity*" (Winery 1) and "*it helps to avoid losing our uniqueness*" (Winery 5). Moreover, residents and the territory benefited from this situation, as some interviewees stated: "*new jobs are created, different skills are required*" (Winery 1). This is consistent with previous studies, where the recognition seems to have positive externalities in terms of creation of new employment and inducing a positive economic impact on the local economy (Caust & Vecco, 2017).

This is a huge impact for such a small destination, where the culture is mostly related to winemaking rather than tourism management and promotion. In fact, one of the main themes that emerged from interviews was the "positive" interference of the COVID-19 pandemic on tourism development. Some wineries stated:

> I have always said that if there was no Covid, we probably would not have been ready to receive the tourist flow we are seeing in recent years. (Winery 7)

When the area earned the recognition, it was not ready to receive the new increased flow of tourists. "Luckily" the pandemic stopped the traveling activities and the community started to improve the tourism offer, but the growth of tourist arrivals unveiled weaknesses in terms of the area's capacity to accommodate and welcome them.

This appears to be not only an infrastructure-related problem, but it also represents a cultural problem.

As far as the first one is concerned, interviews highlighted that accommodation facilities are not sufficient; new ones had to be created and existing ones improved: accommodation capacity showed a limited preparation for the sector with respect to, for example, the possibility of staying overnight in the wineries. Even so the situation is still not ideal, as the offer does not meet the accommodation requests. *"UNESCO brought a rapid increase of tourist arrivals in the area, and we do not have accommodation facilities with large capacity"* (Winery 3).

Moreover, the reception staff is not well trained: new competences and skills are required because the employees are not ready to welcome national and international tourists, or able to satisfy visitors through a catering service and high-level services and information. Another big issue is viability: roads are narrow, the road network is complex, and places of interest are not well indicated.

As far as the second one is concerned, a first element that emerges from the interviews is that the local community and wine producers were not accustomed to large numbers of visitors in the area, and this has influenced how local actors perceive tourists. Wineries 5 and 6 pointed out that some clashes between locals and tourists emerged when tourists visit vineyards:

> Some wine producers believe that their vineyards are theirs and that nobody should enter them. But locals should realize that tourists arrive here because they want to visit our place. (Winery 6)

On the other hand, the feeling is not consistent among all interviewees. Some of them believe that while the local community should be more welcoming towards tourists, the latter should be respectful. The relationship between residents and tourists in the Prosecco region can vary. While some residents may embrace and benefit from tourism development, others may feel a sense of intrusion (or disruption) to their daily lives. This attrition is common in many touristic sites that require efforts to balance the needs of both residents and tourists in order to foster a harmonious relationship. This poses challenges as in the case of overtourism: we refer again to destinations that are facing difficulties due to the rapid growth in tourist arrivals globally.

> Tourists should know that entering private vineyards requires respect, because winegrowers cannot afford to have tourists invading or damaging them. We need to build a trustful relationship between them. (Institutional Actor 1)

This quote emphasizes the importance of managing the transition from a wine-based economy to an enogastronomic tourism economy. The urgency to develop a "culture of hospitality" is acknowledged by interviewees.

> Until now this territory has flourished through wine and vineyards, while from now on we should focus on sustainable tourism development. (Institutional actor 1)

At the same time, there are concerns about the commoditization of products and local practices, as it appears crucial to preserve the place's authenticity.

> We need to be careful because what is great about this area anyway is that the wineries and facilities are all family-run. (Winery 7)

This issue is well known in the extant literature, and it appears to affect not only places that obtained the UNESCO recognition, but it concerns many wider cultural heritage tourism contexts (Zhang & Yin, 2020). When tourism flows have a vast impact on the territory and its economy, the risk of losing a place's authenticity is high.

Tourism Management and Governance

A final important issue concerns the role of institutions managing the UNESCO site in maintaining a fine balance, so that the interests of residents, tourists, and the production system are protected, while concurrently sustainable tourism is developed.

In this case, the promotion of the Prosecco DOCG is linked to the promotion of the Prosecco Hills as a destination, and some informants pointed out that too many small institutions are involved, without a clear orchestration. The main problems emerging from our interviews can be summarized in the following three issues.

First, there are too many institutions that try to manage the Prosecco Hills area. They need to work together, to coordinate flows, activities, and the image of the area, but fragmentation among responsible bodies does not help. Second, a clear and simple accusation towards the communication management as to why there is no information about the UNESCO recognition in the main hubs of the territory (Conegliano and Valdobbiadene), and, finally, why this information is not linked to the DOCG.

> People arrive in the Conegliano or Valdobbiadene main squares and they still don't find any information about the GI. (Winery 7)

Institutional bodies play a crucial part in managing the Prosecco region. Local governments, tourism boards, and other relevant institutions have the responsibility of implementing policies and regulations and might also facilitate communication to ensure the long-term well-being of the community. However, tourism management still relies on Pro Loco voluntary workers (Pro Loco are small associations of local culture and tourism) rather than on skilled professionals with a clear vision to pursue. This is evident, for example, in the lack of basic communication tools, such as GI references. In general, the situation swings between a lack of communication concerning the UNESCO recognition, and the simultaneous involvement of too many institutions clashing one against the other and not generating synergies.

Not all the wineries have a negative attitude about the promotion of Prosecco (wine and hills): some praise the way in which the Consortium is working and how institutions support the UNESCO site as well. On the contrary, others blame the (missing) role of coordination, but highlight that *"UNESCO forces us to grow in a positive way"* (Winery 1), mentioning a network among wineries created thanks to the recognition. This has generated the willingness to team up, creating strong relationships to improve the territory's economy, also to avoid the risk of losing the UNESCO recognition. Thus, common divergences and controversies among the different actors involved in GIs systems (Rinallo & Pitardi, 2019) can be smoothed because the value of the recognition makes the Prosecco GI even more important.

In order to address these critical issues, a number of initiatives have been planned.

The success of the Prosecco area relies also on the engagement of youngsters, consequently it is crucial to raise awareness about the meaning of being members of the World Heritage List. Accordingly, educational activities focusing on heritage preservation and enhancement were included in touristic, culinary, and technical high school programs in the area. Youngsters represent an essential resource for developing new skills lacking in the area:

> We need to increase the sense of belonging to this place among young people, which can support a tourism approach in line with sustainability. We want to work with schools to teach kids about this territory, its landscape, the quality and value of the wine we produce, and the respect they must have for this. (Institutional actor 1)

> We are working on UNESCO awareness programs in schools so that young people can learn what it means to be citizens in a World Heritage Site. (Institutional actor 2)

However, young people are not the only group able to provide new skills for sustainable tourism development. People of all ages employed in tourism, hospitality, and gastronomy need to be upskilled in both how to welcome tourists as well as in terms of how sustainability should be implemented at multiple levels. Therefore, a Wine Tourism Lab has been established as a collaboration between the DOCG Consortium and the local UNESCO Association: *"We need to support this incredible economic opportunity where the landscape has become an economic resource, as people come here to see our landscape. This is also a great responsibility as this is a UNESCO territory as well as a viticultural landscape" (Institutional actor 1).*

These excerpts show that local actors are aware of the need to develop and diffuse a better culture of hospitality, and that this can be achieved through educational/upskilling initiatives directed both at young people and professionals employed in different sub-sectors underlying the new sustainable wine tourism model (hospitality, gastronomy, etc.). Local actors' approach towards education and training for sustainable tourism development is in line with recommendations from the literature that emphasizes the compelling role of education. Indeed, in order to turn sustainable tourism into an accepted social norm, it appears necessary to educate organizations and consumers, as well all interested parties, systematically (Streimikiene et al., 2020). Hence, bringing multiple stakeholders together to participate in dialogue, decision-making, and implementation of responses to this development project.

Change management is a slow motion and process that requires time to avoid a distortion of the whole project. A good relationship between residents and tourists is at the basis of sustainable tourism development: attracting tourists whose values are aligned with those of the host is more important than the quantity of tourists visiting a destination (Dwyer, 2018).

The excerpts discussed above show that the local economy has developed primarily through wine selling (wine-based economy), while the arrival of tourists in the area calls for new approaches to generate a tourism-based economy. As recalled in the literature review, having tourist potential is not sufficient for places to become tourist destinations (Russo & van der Borg, 2002). While data unveils some shortcomings, such as the need to develop accommodation facilities as well as a culture of hospitality, it also shows that local stakeholders are aware of these issues and are introducing several initiatives to address them.

Conclusion

The case at hand, the Prosecco Hills in North East Italy, is a relevant case that puts clearly on display some critical factors entailed in the touristic capitalization of long-held winemaking traditions. The unfolding of the Prosecco Hills Candidacy, and the subsequent mobilization of the area towards the execution of a destination strategy, exposes the problems that might arise from the mismatch between the aspirations of a territory and the lack of tangible and intangible infrastructures to sustain incoming flows. The absence of tangible infrastructures—e.g., bike roads, restaurants—directs our attention towards three sets of critical factors that need to be considered in further research and by practitioners in this and similar areas. These factors refer to three different dimensions: a cognitive one, one entailing the governance of local systems, and the final one related to the development and retention of novel skill profiles.

First and foremost, the area we considered is home for the production of a wine whose domestic and international market shares have grown steadily during the last 30 years. Currently, Prosecco is among the global best-sellers: production volumes topped Champagne ones in the past two years totaling over 750 million bottles. While only partially produced in the area (the Conegliano Valdobbiadene DOCG area is responsible for 100 million bottles in 2022), the wine's striking success triggered a lively entrepreneurial activity and the multiplication of wineries in the GI area, whose focus has been eminently on production processes and marketing. There has been a discussion emphasizing the touristic potential of the Hills for a long time, while in the past 20 years tourism has become visible around the Prosecco Hills thanks to different wine tourism events and the Prosecco Road. However, firms' and operators' attention towards structured and systematic actions aimed at attracting and sustaining incoming flows was quite limited. In other words, the case points to the rigidities affecting firms in "old world" winemaking countries, especially when the market is showing a relentless growth in demand and absorption of the wine. This situation creates incentives for firms—and for their representative bodies such as consortia and trade associations—to focus their attention and investments in operations and marketing and to overlook considerations regarding additional revenue sources such as tourism. When contingencies or institutional actors—in this case policymakers that envisaged the potential of a winemaking area to express its touristic potential—succeed in mobilizing the local actors towards ambitious goals such as the UNESCO recognition, the lack of skills and strategic intent that fit with tourism, or activities that exceed winemaking and marketing can become critical factors.

A second theme emerges from our analysis of the Prosecco Hills case: governing the set-up of the local system of actors that contribute to the deployment of the tourism development strategy. In Conegliano Valdobbiadene, after the successful ignition of the process by the then Ministry of Agriculture, a multiplicity of actors reclaimed a pivotal role in the refinement of the strategy and in its implementation: local associations of municipalities, the Prosecco Consortium, associations of hospitality firms, etc. The presence of several actors with specific competences and prerogatives in a territory is a key element to sustain a collective effort aimed at a successful destination strategy. Nonetheless, the lack of coordination among actors and the absence of criteria to establish the primacy of a leading actor might frustrate such mobilization.

Third, the case stresses the importance of specialized skills for a tourism strategy to take off and for a territory to sustain its attractiveness in time. Being the touristic ambitions relatively recent if compared to the long-held specialization in winemaking and sales, both firms and institutions in the area lack competences and sensitivities to inject novel streams of activities in the extant entrepreneurial fabric. This consideration opens the discussion and further research lines to a second type of attractiveness that territories should take care of to actually benefit from touristic flows: that of skilled tourism actors and novel professional families, alien to the traditional milieu. Such a double effort towards attraction requires to think not only in terms of tangible and intangible infrastructures for tourists but also for novel "residents" or new generations of local dwellers oriented towards career paths that diverge from those strictly connected to winemaking.

References

Carvalho, M., Kastenholz, E., & Carneiro, M. J. (2021). Interaction as a central element of co-creative wine tourism experiences—evidence from Bairrada, a Portuguese wine-producing region. *Sustainability, 13*(16), 9374.

Caust, J., & Vecco, M. (2017). Is UNESCO World heritage recognition a blessing or burden? Evidence from developing Asian countries. *Journal of Cultural Heritage, 27*, 1–9.

Distretto del Conegliano Valdobbiadene Prosecco DOCG. (2022). Rapporto Economico 2022. https://www.prosecco.it/wp-content/uploads/2022/12/Rapporto-Economico-2022.pdf

Dwyer, L. (2018). Saluting while the ship sinks: The necessity for tourism paradigm change. *Journal of Sustainable Tourism, 26*(1), 29–48.

Getz, D., & Brown, G. (2006). Critical success factors for wine tourism regions: A demand analysis. *Tourism Management, 27*(1), 146–158.

Gioia, D. A., Corley, K. G., & Hamilton, A. L. (2013). Seeking qualitative rigor in inductive research: Notes on the gioia methodology. *Organizational Research Methods, 16*(1), 15–31.

Gobbato, E. (2009). *Giuliano bortolomiol: Dreaming of prosecco.* Milano: Veronelli Editore

Patuelli, R., Mussoni, M., & Candela, G. (2013). The effects of world heritage sites on domestic tourism: A spatial interaction model for Italy. *Journal of Geographical Systems, 15*, 369–402.

Rinaldi, C., Finotto, V., Mauracher, C., & Checchinato, F. (2022). *Unpacking the dynamic nature of Geographical Indications (GIs): a longitudinal study on Prosecco as a territorial brand* [Conference presentation abstract]. 6th Annual Conference International Place Branding Association, Aix-en-Provence, France. https://placebranding.org/wp-content/uploads/2022/10/IPBA_Book-of-abstracts_2022.pdf

Rinallo, D., & Pitardi, V. (2019). Open conflict as differentiation strategy in geographical indications: The Bitto Rebels case. *British Food Journal, 121*(12), 3102–3118.

Russo, A. P., & Van Der Borg, J. (2002). Planning considerations for cultural tourism: A case study of four European cities. *Tourism Management, 23*(6), 631–637.

Santos, V. R., Ramos, P., Almeida, N., & Santos-Pavón, E. (2019). Wine and wine tourism experience: A theoretical and conceptual review. *Worldwide Hospitality and Tourism Themes, 11*(6), 718–730.

Sigala, M. (2019). Developing and branding a wine destination through UNESCO World Heritage listing: The case of the Mount Lofty Ranges agrarian landscape. In M. Sigala & R. N. S. Robinson (Eds.), *Wine Tourism Destination Management and Marketing: Theory and Cases* (pp. 113–134). Palgrave Macmillian.

Streimikiene, D., Svagzdiene, B., Jasinskas, E., & Simanavicius, A. (2021). Sustainable tourism development and competitiveness: The systematic literature review. *Sustainable Development, 29*(1), 259–271.

Yang, Y., Xue, L., & Jones, T. E. (2019). Tourism-enhancing effect of World Heritage Sites: Panacea or placebo? A meta-analysis. *Annals of Tourism Research, 75*, 29–41.

Zhang, T., & Yin, P. (2020). Testing the structural relationships of tourism authenticities. *Journal of Destination Marketing and Management, 18*, https://doi.org/10.1016/j.jdmm.2020.100485

Wine Tourism in the Province of Alicante (Spain). Revitalization of Wine Heritage as a Strategy for Tourism Diversification and Differentiation in Mature Tourist Destinations

Rosario Navalón-García

Introduction

In a nutshell, the concept of wine tourism refers to a type of tourism that revolves around the wine experience and its landscape, production, and cultural environment (Blanco-Romero et al., 2021). Other authors such as Hall et al. (2000) define it as an activity comprising the "visitation to vineyards, wineries, wine festivals and wine shows for which grape wine tasting and/or experiencing the attributes of a grape wine region are the prime motivating factors for visitors." In any event, it is an activity in which the sensorial experience relating to wine is the principal motivation, together with the desire to learn about and understand the culture of wine, the vine-growing process and winemaking and its connection with the territory in which it is developed. Therefore, wine tourism is rooted in the cultural, social, economic, and environmental history of

R. Navalón-García (✉)
Department of Regional Geographical Analysis, University of Alicante, Alicante, Spain
e-mail: r.navalon@ua.es

J. Martínez-Falcó et al. (eds.), *Wine Tourism and Sustainability*,
https://doi.org/10.1007/978-3-031-48937-2_9

the producing areas and wineries. In this way, in many wine tourism proposals, the wine landscape (Plaza & Fernández, 2020) (Nogué & San Eugenio, 2017) acquires a leading role, based on the combination of an emerging acknowledgment of its heritage value, arising from tradition and its adaptation as a tourist attraction (López-Guzmán et al., 2013).

Its activation as an element of interest for a growing tourism demand began to be promoted a few decades ago with the double objective of, on the one hand, conserving traditional uses and activities and, on the other hand, fostering endogenous development and economic cohesion in disadvantaged rural areas. In other cases, investment in and the reconversion of farming spaces into wine-producing and wine tourism areas have arisen as a diversification and multifunctionality strategy in contexts of change in agricultural spaces (Cañizares & Ruiz, 2014) in order to adapt to the new trends in the sector.

It is clear that wine tourism is an activity which involves many elements; some related to the agricultural sector but also others related to gastronomy tourism and cultural tourism in a broad sense. Therefore, different approaches may be taken when contemplating wine tourism. On the one hand, it refers to the motivation, enjoyment, and behavior of the consumer with respect to wine and the availability of services that respond to the needs of a changing tourism demand. On the other hand, it may be addressed from the perspective of the strategies of a territory as a key aspect of local development (Martínez & Morales, 2016) and even the actions of the wine market of each area to make themselves attractive in a competitive market. It seems evident that there are many ways of proposing and enjoying wine tourism. This is precisely why it attracts a diverse and broad target market, which is constantly growing. The winemaking regions have responded both in terms of development and product extension strategies and in the creation of new proposals that are of interest for tourists.

Within this context, from the 1990s, there has been a constant growth in the relevance of wine tourism both in rural spaces, traditionally linked to the agricultural activity which is adapted, diversified, and reconverted, and to winemaking areas close to mature tourist spaces that take advantage of the proximity of a demand that seeks variety, so as to develop proposals of change towards wine tourism. At the same time, the transformations taking place in the tourism market, together with the evolution and life cycle of the tourist destinations (Butler, 1980) and the need to adapt the demand receiving spaces to the new global dynamics has led to the implementation of renewal strategies in tourist destinations, but also to the development of initiatives of diversification and differentiation in very different consolidated or emerging spaces (Vera & Baños, 2010). Therefore, both in areas of incipient tourism development in the rural environment and in places close to traditional tourist spaces, the development of wine tourism is related to the needs to diversify the economic structure and is understood as a vector able to give rise to a positive dynamic that generates employment and slows depopulation. This activity also enables a certain differentiation with respect to other producing spaces, in many cases based on the vindication of tradition, the value of the landscape and heritage of the place and the uniqueness and identity of a population (Navalón-García & Rico, 2016). In the case of places close to consolidated tourism dynamics, the wine tourism activity enables new attractions to be generated with the capacity to diversify and differentiate the destinations, also responding to the new motivations and needs of the tourism demand, keen to learn about the places that they visit, sensitive to authenticity, to a variety of proposals and potential experiences.

Within this context of competition between different winemaking spaces and their commitment to wine tourism, this study has the objective of determining whether the supply of the wineries of the Alicante Wine Route adapts to the needs of the demand in terms of its connection of the wine tourism proposals with the landscape and the tangible and intangible heritage of its environment. The results are shown by applying an evaluation matrix developed for this purpose. Based on an analysis of the 19 wineries included in the aforementioned route in 16 municipalities through field work and the consultation of their websites, it may be deduced that the majority of these establishments adapt to the needs of the demand unequally according to the predominant focus of observation: in terms of the different aspects related to the company and their online presence; from the perspective of the proposals related to the heritage and implementation of strategies for interpreting the heritage and history of the place that are able to enhance experiences and link the demand to the local identity; or from the perspective of its relationship with the community of its surroundings.

The article is structured into five sections. After the introduction, a review of the literature related to the object of study is conducted, referring to wine tourism and the market trends, wine routes and the motivations of the demand and some of the ways wineries have adapted to the context of the market competition. Next, the methodology used is described together with the dimensions of the analysis carried out in the wineries of the Alicante Wine Route. The principal results of the study are then reported. Finally the most relevant conclusions are drawn followed by a discussion and the proposal of potential future lines of research arising from this study.

Literature Review

Wine tourism has arisen as a combination of the interests of the winemaking industry and those of tourism (Cassar et al., 2018) and it has enabled wineries and their surrounding territory to become spaces of tourist attraction in different winemaking regions across the world. The so-called *tourist terroir* (Hall et al., 2000), which brings together natural, cultural, and production features in each region to shape a territorial image, a potential experience and a controlled designation of origin, has contributed to the acknowledgment of this activity as an important vector of tourism development. Although there may be a certain tradition in terms of an interest in visiting wineries and learning about wine (Ponce, 2023), the rapid evolution and emergence of wine tourism is due to regional and national strategies (Hall & Sharples, 2008) and the introduction of European policies to support the development of rural areas at the beginning of the 1990s. These policies encouraged the formation and methodology for implementing the wine routes (Millán & Dancausa, 2012), as organized networks that combine the interests of the public and private agents related to wine and its landscapes in order to promote regional development. Within this context, the Secretary of State for Tourism (2000) promoted and defined them as "the integration under the same thematic concept of the existing and potential tourism resources and services of interest of a wine-producing area, consisting of authentic and experience-based proposals in order to construct a product from the very identity of the destination to facilitate the joint commercialisation of the whole area and guarantee the level of satisfaction of the demand, thereby driving its integral economic and social development." The initiative has the double objective of qualifying and diversifying Spanish tourism, while seeking to promote the national and international image of the large Spanish wine companies and positively reinforcing the sales of the wineries, particularly those of small companies with greater difficulties to market their products (Fernández & Vidal, 2020).

The wine routes have the capacity to bring together the interests of the winemaking sector and their territory in order to generate a flow of visitors who wish to discover not only the origin of the product itself but also to learn about the places and production methods. They constitute tools to promote the diversification of regional economies while enhancing the reputation and image of the winemaking regions and their wines (Hall & Mitchell, 2000). In Spain, they constitute the principal formula for organizing wine tourism, based on the coordination of products and itineraries by quality winemaking areas (Miranda & Fernández, 2011). There are currently 35 wine routes in Spain (ACEVIN, 2022) with different functional and organizational structures, adapted to different stages of development and the trajectory of each region.

Due to the diversity of elements and the complexity of the relations that participate in its development, wine tourism and the wine routes can be considered as a system. Its success depends on the smooth functioning of its components (Getz, 2000), so the correspondence between the expectations and needs of consumers with the response of the organizers and public or private agents involved is fundamental in the planning of the wine tourism destinations. In addition to the correct interlinking of the elements, agents, and relations of the system, Singh and Hsiung (2016) report that the critical success factors go beyond a simple associative relationship with wine production; interaction is required between the product, lifestyle, the protection of the heritage and landscape, the brand, the business relations with individuals, and the associations and the participation of the supporting community. In this way, in order to promote wine tourism, wineries, vineyards, and vine growers must act and invest in an integrated way in the continuous improvement of their services and associated infrastructures in order to offer wine tourists an experience that responds to their expectations (Ramos et al., 2018; Santos et al., 2020).

According to Getz and Brown (2006, p. 156), from the point of view of wine tourism consumers, there are three fundamental characteristics that have to be combined: First, the quality of the winemaking products (welcoming wineries for the visitors, qualified personnel, wine-related events); second, the quality and attractiveness of the surrounding landscape; and third, the cultural and tourism context (unique heritage-based accommodation, quality restaurants, architecture and traditions of the location). Therefore, in addition to the supply related to wine and the wine experience, there has to be a wide range of cultural, gastronomic, landscape, and outdoor activities. The wineries are the principal attraction (Getz, 2000), but they cannot work alone. As indicated by Getz and Brown (2006) in their studies on the motivations of long-distance wine tourism, the distance with respect to the outbound tourism markets, adequate accessibility to the places, and the previous knowledge or the existence of tourist packages are aspects that should be taken into account. However, most relevant in the travel decision is the overall attractiveness of the area and the expectations of the diversity of activities and leisure and outdoor proposals in the destination.

The winemaking regions compete with all of the other possible tourist destinations. Therefore, it is fundamental that the organizations in the destination, the wine industry, the community, and the cultural heritage institutions collaborate in creating products and the communication of benefits. It may be true that those responsible for developing wine tourism destinations believe that the key to success resides in the attractiveness of their wines or wineries instead of the overall cultural and recreational experiences sought after by the consumers. However, studies on the image of winemaking regions documented by Williams (2001) show that the winemaking industry has responded to the desires of the tourist demand by incorporating rural and Cultural Landscapes into wine tourism products, as the "wine experience" encompasses a series of elements that go beyond the winery itself.

Following this argument regarding the dimensions of tourism that make up the tourism experience, Moscardo (2020) addresses the issue based on the need of tourists to learn, create, and hear stories about the places they visit. This converts the narrative of places into a central element, not only to promote the destinations, but to respond to the needs of the demand in search of a story and the authenticity of the heritage. In practice, this implies that the destinations and companies should work and improve the efficiency of their discourse to generate opportunities for experiences using the stories of the place, of the winery or the wine itself as a theme and central structure. This will first attract tourists, but also reinforce the brand of the heritage and differentiate the winery from its competitors. The bibliography has identified the heritage, the narration of stories, and authenticity as important concepts for the interaction with tourists in the wineries (Frost et al., 2020). However, it is necessary to understand how this narration of the places and their surroundings is applied by the personal in their interaction with the wine tourists. The narration of stories is considered as a useful strategy to attract tourists and the importance of the authenticity, both intrinsic and existential, according to this author, stands out as means to secure a competitive advantage. Constructing and disseminating the story of the place helps to create tourist experiences. Some of the lines of study in this respect may include; stories about the destination told to the tourists in person by guides or through audiovisual guides; also with text on the signs and in the guides and through digital platforms such as mobile applications; on the other hand, structured activities may be designed to encourage tourists to participate. Both the narration of the place as an in situ activity and the online proposals are relevant in the promotion and attraction potential and reflect the impact of the growing use of the Internet and mobile technologies in tourism as a means of promotion.

There is an increasing trend in the online consultation, planning, and management of trips by tourists and the use of websites by tourist destinations constitutes a strategic management, planning, and marketing tool (Buhalis, 2000; Cassar et al., 2018). They are also a primary means of connecting wineries, organizations, and tourists (Lopes & Rengifo,

2023) and studying them is fundamental for understanding the strategies of development and communication by companies and destinations. It is with good reason that authors have analyzed information technologies and their relationship and impact in wine tourism, to confirm that they improve the coordination and internal and external communication of winemaking companies and their digital promotion, the online sale of products and services and the planning of trips associated with this activity (Alonso et al., 2015; Duarte et al., 2013; Jeong et al., 2012; Torres et al., 2021).

It is acknowledged that the Internet has revolutionized the tourism sector in general and wine tourism is no exception. In fact, as indicated by Cassar et al. (2018), ICTs have given rise to the emergence of websites and platforms that have empowered clients, offering them the opportunity to learn about and select the destination of their trips, make reservations for travel, accommodation, places of interest and services, and even purchase products all online. The small size of many wineries means that they largely depend on the Internet to make themselves visible in the global market, but also to create a reputation and increase the sales of their products. However, despite the importance of the websites, according to Barroco and Amaro (2020), many wine routes and the elements and entities of which they are composed only use a small part of their potential as a communication and management tool. For example, some have unclear, static interfaces with very little information (Camprubí & Galí, 2015).

In light of the above, it is evident that for a website to be effective it must incorporate all those aspects that respond to the needs of the demand, in terms of basic necessary information about the location of the place, access, services, opening hours, and prices. However, in the case of wine tourism, it is also necessary to include the differentiating features, the territorial context and the story of the place, together with a suitable, legible, and agile design as a first contact that indicates the willingness to create the experiences demanded by the potential visitors. Also relevant is the choice of a positioning that enables the differentiation of the products and services of the website.

Methodology

In order to determine whether the development of wine tourism through the Alicante Wine Route is able to invigorate the winemaking heritage as a tourism diversification and differentiation strategy of the Costa Blanca, the supply of the 19 associated wineries located in 16 municipalities of the province has been analyzed. They are located in two main areas. On the one hand, in the valleys of the interior of the districts of Alt and Medio Vinalopó with open landscapes and large properties and a tourism context that has hardly been developed to date. On the other hand, it includes wineries of the Marina Alta district, located in a more mountainous territory with smaller properties located in coastal and pre-coastal municipalities in the north of the province where the presence of international tourists and residents is a distinct feature of their sociodemographic make-up (see Fig. 1).

In order to obtain the necessary information, a qualitative study based on analyzing the contents of the websites of these wineries has been carried out, complemented with field work. So as to be able to offer comparable results, a table has been elaborate that includes the key variables that have already been proposed by different authors, grouped into four dimensions (Lopes & Rengifo-Gallego, 2023; Morrison et al., 2004). In this case, they are adapted to the object of interest in accordance with the needs identified in the demand by the authors consulted in the literature, fundamentally Getz (2000) or Getz and Brown (2006), generating the following evaluation model (Table 1).

- Dimension of the company (D1) which is subdivided into the identifying information of the winery, the virtual information that is provided; the interactivity of the website and the existence of a sales section. Score out of 30.
- Interpretative dimension (D2), which includes a section of information about the story of the winery and its elements. There is also a section on the intellectual accessibility of the website and the existence of interpretative resources and spaces. Score out of 25.
- The wine tourism dimension (D3) is specifically focused on the details of the supply of wine tourism that is provided. Score out of 15.

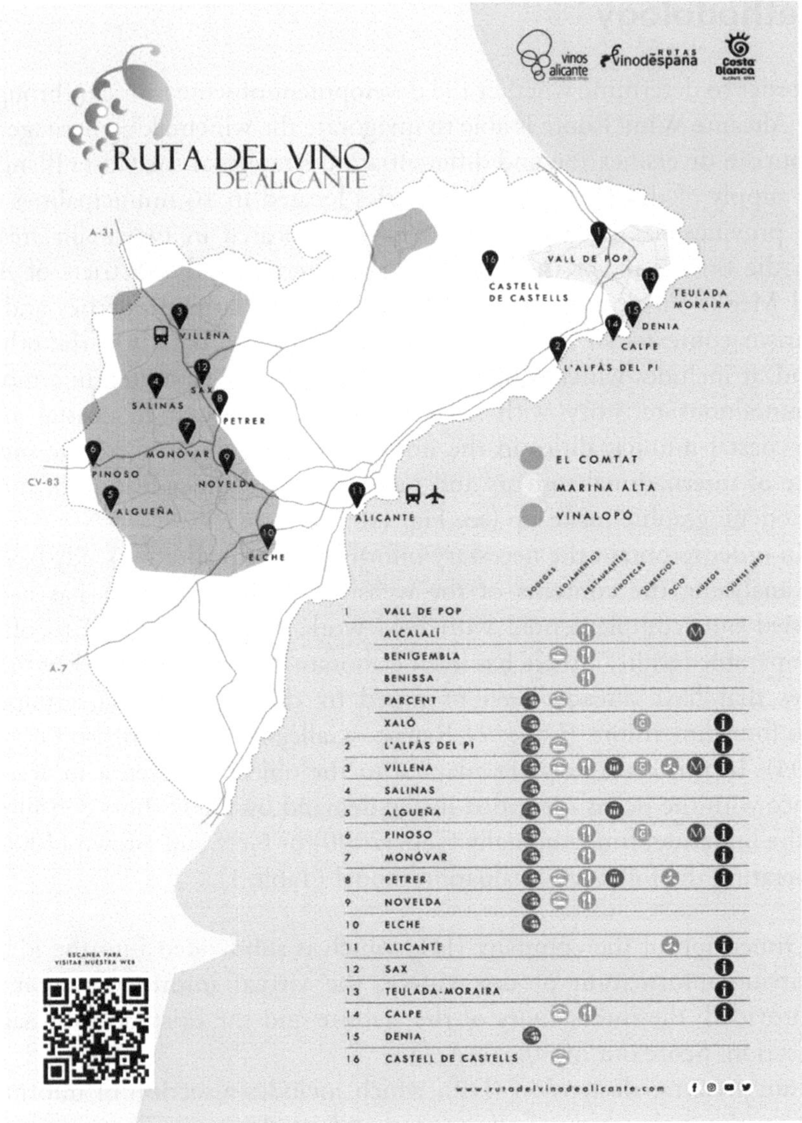

Fig. 1 Map of the Alicante wine route (areas and supply) (*Source* Alicante wine route. Obtained from https://www.calpe.es/sites/default/files/2020-09/MAPA%20RUTA%20DEL%20VINO.pdf)

Table 1 Items analyzed in the study of the wineries

N1	**Name of the winery**		
N2	**Municipality**		
D1	**1. COMPANY DIMENSION**		**score 0–30**
	1.1. Basic information	**Description**	**0–10**
	Name of the company	Is the name of the company well highlighted and clear?	
	Presentation of the company	Is there a presentation of the winery/company?	
	Contact information	Is there an address? [(1); telephone number (2); email address (3); contact person (4)]	
	Opening hours	Are the opening hours shown? (1) Does it open at weekends? (2)	
	Map of the location and GPS coordinates	Is there a map of its location? [(1) and GPS coordinates (2)]	
	1.2. Virtual information		**0–5**
	Leaflets and videos	Is it possible to download leaflets or videos?	
	Virtual tour	Is it possible to make a virtual tour of the winery or company?	
	Links to institutional websites	Are there links to websites of the institutions? (0 = 0)(1 = 1)(2 = 2)(≥3 = 3)]	
	1.3. Interactivity		**0–10**
	News	Is there a news section?	
	Links to news or reviews	Does it have links to websites of journals, newspapers, etc.?	
	Forms	Does it have complaints, request, information forms, etc.?	
	Subscription to newsletter or mailing list	Does it offer the possibility of subscribing to a newsletter?	
	Testimonials of visitors	Does it display testimonials of visitors?	
	Social networks	Does it have links to social networks? [(0 = 0) (1 = 1) (2 = 2) (≥3 = 3)]	

(continued)

Table 1 (continued)

N1	Name of the winery		
	1.4. Sales		**0–5**
	Physical store	Does it have a physical store for selling products?	
	Online store / online sales	Does it have an online store for selling products? (1) does it have a shopping cart? (2)	
	Packs/baskets	Does it offer the possibility of purchasing packs or baskets of several products?	
	Store and/or distributor finder	Does it have a finder for stores or distributors of the company's products?	
D2	**2. INTERPRETATIVE DIMENSION**		**score 0–25**
	2.1. Information of the winery		**0–20**
	History of the winery or company	Is there a description of the history of the winery/ company?	
	Biography of the founders	Is there a biography of the founders of the winery/ company?	
	Presentation and bios of collaborators	Is the team of collaborators presented? [(1)and their biography (2)]	
	Information on the ownership	Is there a description of the ownership?	
	Information about the vineyards	Is there a description of the vineyards?	
	Information about the wines	Is there a description of the wines produced?	
	History - storytelling of the wines	Is the history of the wines presented? (1) Is there any storytelling about the wines presented? (2)	
	Information about other products	Is there a description of any products other than wine?	
	History - storytelling of other products	Are the products of the place connected with the wines?	

(continued)

Table 1 (continued)

N1	Name of the winery		
	Infrastructures (building)	Are the infrastructures of the winery or company shown?	
	History of the building	Is there any information about the building? (1); architecture (2); history	
	2.2. Interpretative services and resources		**0–5**
	Accessibility	Does it indicate whether the winery is accessible for people with disabilities? (1) Is the website adaptable? (2)	
	Exhibition space	Is there an exhibition space about the winery?	
	Interpretative activities	Are there activities adapted for different groups?	
D3	**3. WINE TOURISM DIMENSION**		**score 0–15**
	3.1. Wine tourism		**0–15**
	Tourism services proposals	Does it present itself as a wine tourism experience?	
	Wine tourism packages or products	Does it offer tourist packages? [(0 = 0) (1 = 1) (2 = 2) (≥3 = 3)]	
	Description of activities	Is there any information about each of the activities developed?	
	Human resources or tourism services	Are the human resources related to wine tourism presented? (guides = 1) (oenologist = 2)	
	Visual information (photos, videos)	Does it show (0) photos (1) and videos (2)?	
	Opening times and agenda	Is there a schedule of activities?	
	Prices of the activities	Is there a price list of the activities?	
	Information for individual visitors	Is there any information for individual visitors?	
	Information for groups/ families	Is there any information for groups or families?	

(continued)

Table 1 (continued)

N1	Name of the winery		
	Complementary activities	Are there any complementary activities?	
	Online reservations	Is it possible to make online reservations for wine tourism activities?	
D4	**4. COMMUNITY DIMENSION**		**score 0–10**
	4.1. Relationship with tourism in the area		**0–10**
	Link to the Wine Route website	Does it have a link to the Wine Route website?	
	Links to the websites of partners of the route	Does it have links to the websites of partners?	
	Information about the DO/CVR	Does it provide information about the DO/CVR and a link to the institutional page?	
	Information about the tourist destination	Does it provide information about the tourist destination in which it is located?	
	Links to tourism entities	Does it have links to the websites of tourism entities and institutions in the area?	
	Links to accommodation websites	Does it have connections with local accommodation enterprises?	
	Links to restaurants	Does it have connections with local restaurants?	
	Links to other tourism resources	Does it have links to other regional tourism resources (museums, etc.)?	
	Participation in the local community	Does the winery actively participate in the local community?	
	Environmental and sustainability actions	Are activities related to environmental sustainability promoted?	
	D1 + D2 + D3 + D4 =		**score**

Source Own elaboration

– The community dimension (D4) refers to the elements in the website that indicate the existence of a relationship between the winery and its surroundings, with the Alicante Wine Route or others, with the Designation of Origin or with the tourist attractions. Score out of 10.

The potential sum of 80 points will serve to evaluate the degree of adaptation of the wineries to the needs of the wine tourism demand, with special emphasis on those aspects that connect the wineries with their heritage or the context; the use of actions to link the values of the place with techniques of interpreting the heritage and with the history and stories about the surroundings; and the elements of the wine, heritage, or tourism offer of the surroundings that would increase the interest of a potential demand.

In addition to this analysis of the websites of the wineries, the wine experiences offered through the Alicante Wine Route website have also be consulted up to the month of May 2023. The principal objective is to determine whether these specific offers, which combine different services in a single proposal, adapt to the motivations of the demand which prefers a diverse range of activities in the winery and in the outdoors (Afonso et al., 2018), and, in their choice of trip look for wine and local gastronomy tasting but also value the options of discovering the heritage and landscape of the places that they visit and of being busy during their stay (Getz & Brown, 2006).

Results

The balance of the results after consulting the websites and undertaking the field work reveals an uneven result in each of the dimensions studied. Although the overall percentage balance indicates that the average global score is below half of the possible total (40.5%), there are dimensions that are clearly above this level, reaching 73.68% in the case of the basic information about location and the identification of the wineries. Others, however, barely reach 14.7%, as in the case of the downloadable virtual information.

With respect to the development of each of the dimensions, as shown in Table 2, the dimension referring to the information of the company (D1) is the most developed in merely informative aspects, but also in the section of sales, which is reached in 58,9% of cases. There is room for improvement in aspects related to the virtual information of the wineries. Furthermore, everything related to the interaction with visitors and testimonials is clearly revisable, with barely 10.5% of wineries complying. Despite the creation and co-creation of stories in the experience of the company that forms a connection with the demand, this aspect is one that could be reinforced in the satisfaction of wine tourism demand (Frost et al., 2020).

With respect to the interpretative dimension (D2) (Table 3) in which the degree of the incorporation of content related to the history of the

Table 2 Percentage of fulfillment in the dimension of analysis of the company (D1)

D1	**1. COMPANY DIMENSION**	**47,5%**
D1.1	**1.1. Basic information**	73,68
D1.1.1	Name of the company	100,00
D1.1.2	Presentation of the company	94,74
D1.1.3	Contact information	50,00
D1.1.4	Opening hours	76,32
D1.1.5	Map of the location and GPS coordinates	94,74
D1.2	**1.2. Virtual information**	14,74
D1.2.1	Leaflets and videos	26,32
D1.2.2	Virtual tour	0,00
D1.2.3	Links to institutional websites	15,79
D1.3	**1.3. Interactivity**	42,63
D1.3.1	News	52,63
D1.3.2	Links to news or reviews	21,05
D.1.3.3	Forms	94,74
D1.3.4	Subscription to newsletter or mailing list	21,05
D1.3.5	Testimonials of visitors	10,53
D1.3.6	Social networks	68,42
D.1.4	**1.4. Sales**	58,95
D1.4.1	Physical store	84,21
D1.4.2	Online store / online sales	76,32
D1.4.3	Packs/baskets	47,37
D1.4.4	Store and/or distributor finder	10,53

Source Own elaboration

winery, the ownership, the buildings, and the people is analyzed, the overall score is low, with only 34.7%. However, in the majority of cases, information about the ownership, the wines, and even the building of the winery is included, but not so much the story of these aspects that would enable a connection to be created based on its narration. The figure that clearly lowers the average for this dimension refers to the accessibility to the wineries by people with disabilities, as they are not adapted for them. This affects the overall calculation, together with the low presence of exhibition spaces in the wineries. On the other hand, there is a supply of activities nearby related to the interpretation of the heritage in more than 58% of cases.

The results referring to the wine tourism dimension (Table 4) of the offer directly related to the provision of services to tourists in the wineries are positive, with an overall figure of compliance of over 57%. Certain items are clearly achieved with the existence of a supply of services, while others, such as human resources specialized in the service barely reach 18.4%, despite being an issue that is clearly valued in the studies carried out by Getz and Brown (2006). Other aspects that can be clearly

Table 3 Percentage of fulfillment in the dimension of analysis of heritage interpretation (D2)

D2	**2. INTERPRETATIVE DIMENSION**	34,73%
D2.1	**2.1. Information of the winery**	55,79
D2.1.1	History of the winery or company	78,95
D2.1.2	Biography of the founders	68,42
D2.1.3	Presentation and bios of collaborators	36,84
D2.1.4	Information on the ownership	100,00
D2.1.5	Information about the vineyards	71,05
D2.1.6	Information about the wines	100,00
D2.1.7	History - storytelling of the wines	65,79
D2.1.8	Information about other products	36,84
D2.1.9	History - storytelling of other products	42,11
D2.1.10	Infrastructures (building)	89,47
D2.1.11	History of the building	35,09
D2.2	**2.2. Interpretative services and resources**	13,68
D2.2.1	Accessibility	0,00
D2.2.2	Exhibition space	5,26
D2.2.3	Interpretative activities	57,89

Source Own elaboration

Table 4 Percentage of fulfillment in the dimension of wine tourism (D3)

D3	**3. WINE TOURISM DIMENSION**	**57,19**
D3.1	**3.1. Wine tourism**	57,19
D3.1.1	Tourism services proposals	89,47
D3.1.2	Wine tourism packages or products	46,05
D3.1.3	Description of activities	78,95
D3.1.4	Human resources or tourism services	18,42
D3.1.5	Visual information (photos, videos)	68,42
D3.1.5	Opening times and agenda	36,84
D3.1.6	Prices of the activities	63,16
D3.1.7	Information for individual visitors	57,89
D3.1.8	Information for groups/families	73,68
D3.1.9	Complementary activities	31,58
D3.1.10	Online reservations	68,42

Source Own elaboration

improved are those related to the proposal of other complementary activities and clearly specified opening times and calendars.

The so-called community dimension (D4) (Table 5) seeks .to determine the degree of implication of the winery in its territorial context through the links that are shown on its website to entities and bodies with which it should have close ties, such as the Alicante Wine Route or the DO Vinos de Alicante, for example. No less relevant is its collaboration with the tourism agents of the surrounding area and establishments with which it may have service agreements. The overall score of this dimension is discreet, with an average degree of compliance of barely 22.63%, particularly when taking into account that the demand partly selects the visits based on the possibility of finding other activities (related to wine tourism or otherwise) during their stay in the place. It should be noted, however, that there are ties with environmental and conservation actions in the area and with the local community.

In order to complement the aspects referring to the wine tourism offer of the Alicante Wine Route described in Dimension 3, each of the 38 wine tourism proposals were analyzed to determine in detail which activities were incorporated and whether they connected with the needs of the demand. The results, which are summarized in Fig. 2, show that practically all of the proposals included a visit to the winery and 100% incorporated a wine tasting activity. However, only a small percentage

Table 5 Percentage of fulfillment in the Community dimension (D4)

D4	**4. COMMUNITY DIMENSION**	22,63%
D4.1	**4.1. Relationship with tourism in the area**	22,63
D4.1.1	Link to the Wine Route website	31,58
D4.1.2	Links to the websites of partners of the route	26,32
D4.1.3	Information about the DO/CVR	10,53
D4.1.4	Information about the tourist destination	21,05
D4.1.5	Links to tourism entities	5,26
D4.1.6	Links to accommodation websites	0,00
D4.1.7	Links to restaurants	15,79
D4.1.8	Links to other tourism resources	5,26
D4.1.9	Participation in the local community	52,63
D4.1.10	Environmental and sustainability actions	57,89

Source Own elaboration

include, for example, a visit to the vineyard, reducing the possibility of generating an emotional connection with the territory, the wineries, and the winemaking process. Meanwhile, the number of wineries offering the tasting of other products of the area (cured meats, cheeses, etc.) could be improved, together with those that incorporate the matching of wines with gastronomic proposals in nearby restaurants. There is a very low percentage of offers that include overnight stays. It should be highlighted that, although only symbolically, in some cases the possibility of visiting the cultural and historical heritage of the municipality in which the winery is located is incorporated and in some cases "a la carte" experiences or workshops are offered.

It seems evident that the scarce incorporation of the elements of the territorial and heritage context and the possibilities offered by the tourism services and companies in the area reduce the possibility for wine tourism to become a true vector of economic diversification and of the tourism image of destinations.

Conclusions

Both in consolidated and mature tourism contexts and in emerging spaces, or those undergoing incipient development, the potential diversifying and differentiating element of the activities related to wine tourism

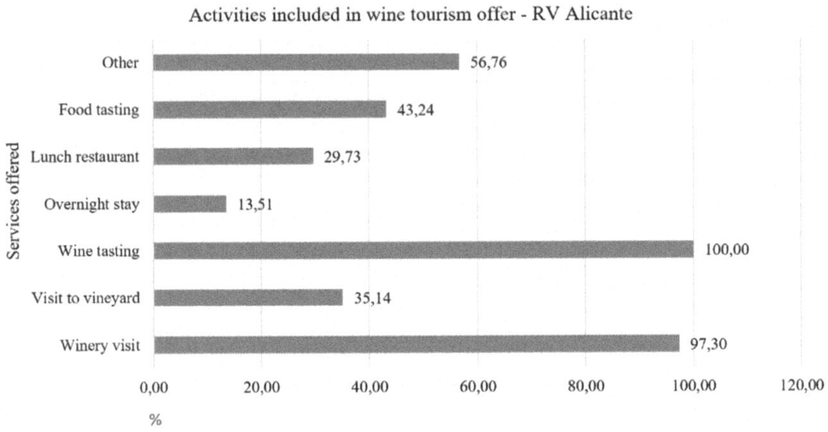

Fig. 2 Wine tourism activities included in the Alicante wine route (*Source* Own elaboration)

is evident. The development of this activity requires the implementation of initiatives coordinated between the public and private agents related to the production and marketing of wine in a space. The creation of wine tourism or wine routes formulate the integrating vocation of these initiatives in the territory both with respect to the landscape and social dimension and its transversal economic invigoration. Therefore, it seems clear that contemplating a wine tourism initiative that implies the structuring of a space around a specific theme for tourism purposes could have the effect of incorporating new assets into the tourism activity. Hence, a well-designed structure and planning will give rise to the renovation of this space and a diversification of the tourism image and activity. This effect also occurs in a non-tourist space to which wine tourism contributes new lines of development and economic and social invigoration. If the landscape heritage and winemaking industry elements are used appropriately, it is also possible to develop strategies of differentiation with respect to other places with similar features.

The potential diversifying, invigorating, and differentiating role of wine tourism is evident if it is planned and implemented in a coordinated and efficient way, generating positive synergies in all elements of the system. Taking this context into account and the global trends in tourist demand and that of wine tourists who seek experiences, variety and proposals able to generate unique memories, it seems clear that the tourist destinations have to act in a coordinated way. On the other hand, the incorporation of ICTs and the use of the Internet to promote and market the tourism products raises the need for the tourist supply to respond to the expectations and needs of the demand in an anticipated way. For this reason, confirming the degree of adaptation of the websites of the wineries of the Alicante Wine Route, together with field work, allows us to obtain a realistic picture of the correspondence between the potential and real development of the actions that are currently being carried out.

The analysis conducted shows that, although there is a growing supply with a good distribution in the two winemaking areas of Alicante, there are still clear lines for improvement. While practically all of the wineries provide basic information about their location, the characteristics of the wines and features of the property or winery, other aspects are not sufficiently addressed and it is precisely these factors that connect with the expectations of the demand, according to the bibliography consulted. The possibility of visiting the vineyards, participating in different activities in the surrounding areas of the wineries or learning about the story and history of the place are offered by a very small percentage of the wineries analyzed. On the other hand, coinciding with the results obtained by other authors, it may be observed that the websites are static as the items related to the online interaction with visitors can clearly be improved. The coordination and connection with other tourism companies and resources in the area is another of the weaknesses detected, with the resulting loss of opportunities to generate territorial synergies. All of this implies that, although the potential is evident, the development of the activities that would enable an invigoration of certain agricultural and/or tourist areas based on the diversification of activities has yet to be consolidated.

With respect to the differentiation potential, the fact that all of the wineries propose very similar activities, namely wine tasting and visiting the winery, gives rise to the need to focus more on those aspects that will make a wine tourism proposal unique, such as the coordination with activities that connect with the local heritage and which draw from the uniqueness of the stories of the place and the family history; even the proposal of activities in the surrounding landscape related to the winery or actions in which the participation of the visitors enables them to generate unique memories.

Acknowledgements Grant PID2020-114186RB-C21 funded by MCIN-AEI/ https://doi.org/10.13039/501100011033

Bibliography

ACEVIN (2023) Rutas del Vino de España. https://wineroutesofspain.com/

Afonso, C., Silva, G., Gonçalves, H., & Duarte, M. (2018). The role of motivations and involvement in wine tourists' intention to return: SEM and fsQCA findings. *Journal of Business Research, 89*, 313–321.

Alonso, A. D., Bressan, A., O'Shea, M., & Krajsic, V. (2015). Perceived benefits and challenges to wine tourism involvement: An international perspective. *International Journal of Tourism Research, 17*(1), 66–81. https://doi.org/10.1002/jtr.1967

Barroco, C., & Amaro, S. (2020). Examining the progress of the Dão wine route wineries' websites. *Journal of Tourism and Development, 33*, 29–40.

Blanco-Romero, M. A. et al. (2021). *Diccionario de turismo cátedra.* ISBN 978-84-376-4200-0.

Buhalis, D. (2000). Tourism and information technologies: Past, present and future. *Tourism Recreation Research, 25*(1), 41–58. https://doi.org/10.1080/02508281.2000.11014899

Butler, R. W. (1980). The concept of a tourist area cycle of evolution: Implications for management of resources The Canadian Geographer. *XXIV, I*, 5–12.

Camprubí, R., & Galí, N. (2015). An exploratory analysis of wineries websites functionality: The case of the DOQ Priorat Route wineries. Boletin de La Asociacion de Geografos Españoles 68, 159–176 https://doi.org/10.21138/bage.1857

Cañizares, M. C., & Ruiz, A. R. (2014). Evolución del paisaje del viñedo en Castilla-La Mancha y revalorización del patrimonio agrario en el contexto de la modernización. Scripta Nova. Revista de Geografía y Ciencias Sociales, XVIII (498), 1–18. https://www.ub.edu/geocrit/sn/sn-498.htm

Cassar, M. L., Caruana, A., & Konietzny, J. (2018). Positioning narratives of wine tourism websites: A lexical analysis across two different regions. *Journal of Wine Research, 29*(1), 49–63. https://doi.org/10.1080/09571264.2018.1433140

Duarte, A., Bressan, A., O'Shea, M., & Krajsic, V. (2013). Website and social media usage: Implications for the further development of wine tourism, hospitality, and the wine sector. *Tourism Planning and Development, 10*(3), 229–248. https://doi.org/10.1080/21568316.2012.747989

Fernández, J., & Vidal, M. J. (2020). Wine routes as engines of socio-territorial dynamisation: The case of Castilla y León. Boletín de la Asociación de Geógrafos Españoles (84) 2789, 1–36. https://doi.org/10.21138/bage.2789

Frost, W., Frost, J., Strickland, P., & Smith Maguire, J. (2020). Seeking a competitive advantage in wine tourism: Heritage and storytelling at the cellar-door. *International Journal of Hospitality Management, 87*, 102460. https://doi.org/10.1016/j.ijhm.2020.102460

Getz, D. (2000). Explore wine tourism. *Management, development and destinations.* Cognizant Communication Corporation.

Getz, D., & Brown, G. (2006). Critical success factors for wine tourism regions: A demand analysis. *Tourism Management, 27*(1), 146–158.

Hall, C. M., Sharples, E., Cambourne, B., & Macionis, N. (eds.) (2000). *Wine tourism around the world: Development, management and markets*, Butterworth-Heinemann, p. 348. ISBN 0 7506 4530 X (Hbk).

Hall, C. M., & Mitchell, R. (2008). *Wine marketing: A practical guide*. Elsevier Ltd.

Jeong, C., Holland, S., Jun, S. H., & Gibson, H. (2012). Enhancing destination image through travel website information. *International Journal of Tourism Research, 14*(1), 16–27. https://doi.org/10.1002/jtr.827

Lopes, C., & Rengifo-Gallego, J. I. (2023). El enoturismo en Internet. Evaluación de los sitios web de las bodegas de tres rutas del vino de Extremadura (España) y de Alentejo y Região Centro (Portugal) Investigaciones Turísticas (26), 207–234. https://doi.org/10.14198/INTURI.21870

López-Guzmán, T., Rodríguez-García, J., & Vieira, A. (2013). *Revisión De La Literatura Científica Sobre Enoturismo En España Cuadernos De Turismo, 32*, 171–188.

Martínez, A., & Morales, F. J. (2016). El vino como recurso turístico para el fomento del desarrollo local: una oportunidad para las comarcas del Vinalopó (Alicante) y el Altiplano Yecla-Jumilla (Murcia) Cuadernos de Turismo 38, 267–300. https://doi.org/10.6018/turismo.38.271451

Millán, G., & Dancausa, M. G. (2012). El desarrollo turístico de zonas rurales en España a partir de la creación de rutas del vino: un análisis DAFO. Teoría y praxis 12, 52–79. https://doi.org/10.22403/UQROOMX/TYP12/03

Miranda, B., & Fernández, R. (2011) Vino, turismo e innovación: las Rutas del Vino de España, una estrategia integrada de desarrollo rural. *Estudios de Economía Aplicada, 29*(1), 129–164. http://revistas.udc.es/index.php/rotur/article/view/1260/338

Mitchell, R., & Hall, C. M. (2006). Wine tourism research: The state of play. *Tourism Review International, 9*, 307–332.

Morrison, A. M., Taylor, J. S., & Douglas, A. (2004). Website evaluation in tourism and hospitality. *Journal of Travel & Tourism Marketing, 17*(2–3), 233–251. https://doi.org/10.1300/J073v17n02_18

Moscardo, G. (2020). Stories and design in tourism. *Annals of Tourism Research, 83*, 102950. https://doi.org/10.1016/j.annals.2020.102950

Navalón-García, R., & Rico, E. (2016). El desarrollo de productos de turismo cultural. *Canelobre, 66*, 278–289.

Nogué, J., de San Eugenio, J. (2017). La contribución del paisaje visual en la generación de marcas territoriales. *Boletín de la Asociación de Geógrafos Españoles* 74, 143–160. https://doi.org/10.21138/bage.2448

Plaza, J. I., & Fernández, R. (2020). Valor patrimonial de los paisajes del vino y su promoción en los valles y llanuras del Este de Zamora. Estudios Geográficos 81:289, julio-diciembre 2020 e046. https://doi.org/10.3989/estgeogr.202060.060

Ponce, G. (Ed.) (2023). Los vinos valencianos. Historia, innovación y paisaje. Col. L'Ordit. Universidad de Alicante.

Ramos, P., Santos, V., & Almeida, N. (2018). Main challenges, trends and opportunities for wine tourism in Portugal. *Worldwide Hospitality and Tourism Themes, 10*(6), 680–687. https://doi.org/10.1108/WHATT-08-2018-0055

Santos, V., Ramos, P., Almeida, N., & Santos-Pavón, E. (2020). Wine experience scale development: A new strategy to measure holistic behaviour of wine tourists. *Sustainability, 12*(19), 8055. https://doi.org/10.3390/su12198055

Secretaría de Estado de Turismo. (2000). *Propuesta para la asistencia técnica para el desarrollo del estudio de definición del producto Rutas del Vino de España.* Servicio de Publicaciones del Ministerio de Industria, Comercio y Turismo.

Singh, N., & Hsiung, Y. (2016). Exploring critical success factors for napa's wine tourism industry from a demand perspective. *Anatolia - an International Journal of Tourism and Hospitality Research, 27*(4), 433–443.

Torres, J. P., Barrera, J. I., Kunc, M., & Charters, S. (2021). The dynamics of wine tourism adoption in Chile. *Journal of Business Research, 127*, 474–485. https://doi.org/10.1016/j.jbusres.2020.06.043329-353

Vera, J. F., & Baños, C. J. (2010). Renovación y reestructuración de los destinos turísticos consolidados del litoral: Las prácticas recreativas en la evolución del espacio turístico. *Boletín De La Asociación De Geógrafos Españoles, 53*, 329–353. ISSN 0212-9426

Williams, P. (2001). Positioning wine tourism destinations: An image analysis. *International Journal of Wine Marketing, 13*(3), 42–59.

Epilogue

The Impact of Wine Tourism on the Sustainability Outcomes of Spanish Wineries: An Exploration via Structural Equation Modeling

Javier Martínez-Falcó, Bartolomé Marco-Lajara, Eduardo Sánchez-García, and Luis A. Millán-Tudela

Introduction

The recent evolution in tourist demand has markedly reshaped the tourism sector, prompting shorter, frequent trips to diverse destinations beyond traditional sun-soaked beaches (Toubes et al., 2021). Consequently, novel tourist products have emerged to cater to these shifting preferences. Notably, gastronomy, with an emphasis on wine, is surging

J. Martínez-Falcó (✉) · B. Marco-Lajara · E. Sánchez-García · L. A. Millán-Tudela
University of Alicante, Alicante, Spain
e-mail: javier.falco@ua.es

B. Marco-Lajara
e-mail: bartolome.marco@ua.es

E. Sánchez-García
e-mail: eduardo.sanchez@ua.es

L. A. Millán-Tudela
e-mail: luisantonio.millan@ua.es

J. Martínez-Falcó et al. (eds.), *Wine Tourism and Sustainability*, https://doi.org/10.1007/978-3-031-48937-2_10

as a primary travel motivation, rather than a mere adjunct (Serra-Cantallops et al., 2021). Such a synergy between wine and tourism not only satiates tourists' quest for unique experiences but also bolsters regional development (Anderson, 2013).

Wine tourism, as delineated by Hall and Macionis (1998, p. 267), traditionally encompasses visits to vineyards and wineries driven primarily by wine tasting and regional experiences. Nonetheless, this notion offers a multifaceted canvas for interpretation. In this sense, Getz (2000) offers a triadic analysis: for wineries, it is a strategic differentiation and a direct sales channel; for tourist regions, it enhances territorial value and conserves vinicultural traditions; and for tourists, it fulfills the pursuit of novel experiences. This study conceptualizes wine tourism as an endeavor intertwined with vineyard appreciation, winery competitiveness enhancement, territorial development, and market innovation.

Prominent instances of wine tourism include regions such as Piedmont, Bordeaux, Oporto, La Rioja, Mendoza, and Napa Valley. While New World wine nations have historically embraced this activity, it gains momentum in Old World countries as a strategic response to changing domestic wine consumption, export inclinations, and evolving tourist preferences for succinct, novel experiences (Gómez et al., 2019). This study zeroes in on Spain's burgeoning wine tourism landscape, which augments the nation's economic, social, and environmental prosperity (Zamarreño-Aramendia et al., 2021). According to recent data from Spain's Wine Route Tourist Observatory (OTRVE), the nation's 35 wine routes attracted over 1.6 million tourists in 2021, yielding an economic impact surpassing 51 million euros (OTRVE, 2021).

Wine tourism significantly impacts regional development by fostering economic and social welfare and ensuring environmental preservation (Festa et al., 2020). It aligns with the Triple Bottom Line's facets: economic, social, and environmental dimensions (Sun & Drakeman, 2022). This research seeks to ascertain wine tourism's influence on the sustainable performance of Spanish wineries. Despite extensive investigations into wine tourism's contributions to territorial development across these three facets, its specific impact on the sustainable operations of Spanish wineries remains uncharted. Thus, diverging from broader studies, this research adopts a nuanced, winery-centric lens to explore this interplay.

To fulfill the research aim, following this introduction, Section 2 delves into the connection between wine tourism and wineries' sustainable performance. Section 3 outlines the methodology employed. Section4 showcases the evaluation results for the overall, measurement, and structural models. Section5 concludes with key findings, study limitations, and prospective research directions.

Wine Tourism and Sustainable Performance

Initial wine tourism research, emerging in the final decade of the twentieth century, primarily examined its nexus with rural economies and wine tourist behaviors (Hall, 1996). Noteworthy seminal works include *Wine Tourism Around the World* (Hall et al., 2000) and *Explore Wine Tourism* (Getz, 2000). Other pioneering contributions by researchers like Charters and Ali-Knight (2002), Bruwer (2003), Alant and Bruwer (2004), Carlsen (2004), Getz and Brown (2006), and Mitchell and Hall (2006) were particularly groundbreaking.

In recent years, wine tourism research has diversified, encompassing themes from marketing strategies to its socio-cultural relevance (Kotur, 2023). Notably, the interplay between wine tourism and sustainability has gained prominence, with the sector promoting environmental stewardship, fostering demand for sustainable offerings, augmenting income streams, and heightening visitor environmental consciousness (Lorey et al., 2023).

Rising temperatures in recent times have jeopardized the continuity of viticultural regions, transitioning sustainable practices from mere competitive differentiation to a strategic imperative for the sector's longevity (Frigon et al., 2020). Wine tourism can bolster sustainability pillars within wineries by creating jobs, promoting heritage awareness, and enhancing environmental cognizance among tourists (Frost et al., 2020). Beyond mere sales augmentation, wine tourism should be viewed as a conduit for broader societal and environmental benefits, crucial for the enduring success of the wine industry (Andrade-Suárez & Caamaño-Franco, 2020). In this line of thought, González (2017) posits that the unique authenticity of each viticultural region can be leveraged through wine tourism, enhancing not only the economic, but also the social and environmental dimensions of wineries. Consequently, wineries engaging in wine tourism can concurrently bolster their competitiveness and sustainability (Nave et al., 2021).

Economically, wine tourism serves as a direct sales conduit for wineries (Sun & Drakeman, 2022). Direct in-winery sales, a dominant metric for assessing the activity's economic footprint (Smyczek et al., 2020), offer benefits such as: (1) competitive pricing or enhanced margins by bypassing intermediaries, (2) immediate cash flow compared to delayed payments in other channels, (3) opportunities for up-selling or cross-selling, and (4) fostering direct customer relations, pivotal for ensuring future sales through brand loyalty, and increased wine variety awareness. Therefore, wine tourism can bolster wineries' economic outcomes via in-winery direct sales and enhanced commercial capabilities (Canziani et al., 2017). Particularly for smaller wineries, which often face challenges accessing broader distribution networks, wine tourism emerges as a lucrative and feasible avenue for wine marketing (Karagiannis & Metaxas, 2020).

Socially, wine tourism addresses depopulation challenges and elevates societal and worker well-being. The economic gains and diversified roles it fosters enhance job security and conditions for winery staff (Williams, 2021). It also paves the way for broader local employment prospects and equips individuals with skills relevant to the burgeoning service sector associated with wineries (Kutateladze et al., 2021). Beyond the winery confines, the community reaps benefits like urban enhancements and enriched cultural and leisure activities (Duarte-Alonso et al., 2021). Consequently, wine tourism can profoundly shape the territory surrounding wineries, financing vital infrastructures like roads and sewage systems, boosting related sectors such as cultural tourism, and facilitating connections with local vendors. It also enriches the locale with additional services and amenities for inhabitants, such as recreational venues and outdoor activities. Furthermore, this form of tourism champions viticultural landscapes and the regional wine culture, safeguarding both tangible and intangible heritages rooted in the area's vinicultural legacy (Carvalho et al., 2021).

From an environmental perspective, wine tourism accelerates wineries' ecological sustainability. This can be driven by a winery's internal motivation for differentiation or external pressures to enhance the surrounding environment. Such initiatives encompass heritage conservation, gastronomic richness enhancement, and biodiversity protection (Darnay, 2016). Consequently, wine tourism fosters eco-friendly practices in wineries, positioning them as hubs for environmentally-conscious tourists (Montella, 2017). Likewise, wine tourism initiatives can enhance ecological understanding. Staff interactions with experts like oenologists and agronomists, coupled with training on sustainable winery practices, bolster their capacity for ecological innovation, thus elevating green performance (Karagiannis & Metaxas, 2020).

Wine tourism, envisioned as the safeguarding and utilization of viticultural heritage, should prioritize cultural and environmental conservation while enhancing the socio-economic well-being of the hosting wine region. Despite its evident role in fostering holistic development, limited scholarly work has assessed its impact on the trifecta of winery performance: economic, social, and environmental. Stemming from the literature review, this study introduces three research hypotheses:

H1 Wine tourism positively influences wineries' economic outcomes.
H2 Wine tourism positively influences wineries' social outcomes.
H3 Wine tourism positively influences wineries' green outcomes.

Figure 1 depicts the proposed theoretical model under examination.

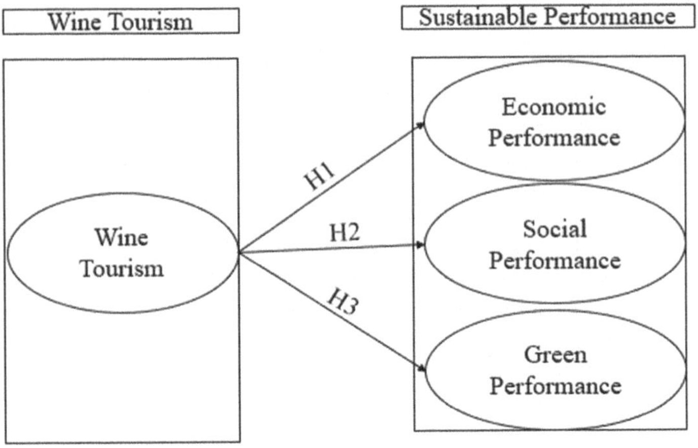

Fig. 1 Theoretical model to be tested (*Source* own elaboration)

Methodology

For a comprehensive understanding of the methodological dimensions pertinent to this study, this section has been delineated into four integral segments: (1) contextual foundation of the research, (2) methodology for population and sample acquisition, (3) procedures for variable quantification, and (4) technique of analysis.

Research Context

Several rationales underscore the selection of the Spanish wine context for this investigation. Initially, based on the latest figures from the International Organization of Vine and Wine (OIV), Spain has been identified as an industry benchmark, renowned for its premier global position in vineyard acreage and wine export volume. Yet, the significance of the wine sector to Spain transcends merely economic metrics. Socio-environmental factors are pivotal as the industry augments citizen well-being and bolsters the preservation of both heritage and the encompassing environment (Marco-Lajara et al., 2022a). Subsequently, a marked ascent in the prominence of wine tourism for Spanish wineries has been observed. Following the inception of the "Wine Routes of Spain" tourism initiative in 2008, a surge in wineries embracing and championing this tourism niche has occurred, culminating in 35 distinct wine routes within the Iberian landscape (Vázquez-Vicente et al., 2021). Lastly, it has been discerned that wine tourism embodies an increasingly strategic facet in buttressing the longevity of Spanish wineries. Not only does it serve as a conduit for direct in-winery sales, offsetting diminishing domestic wine consumption, but it also catalyzes the emergence of fervent brand advocates (Zamarreño-Aramendia et al., 2021).

Population and Sample

The research cohort was drawn from enterprises encompassed within the National Classification of Economic Activities (CNAE, for its acronym in Spanish) under the 1102 code related to manufacturing. As per the Iberian Balance Sheet Analysis System (SABI, for its acronym in Spanish) database, this comprises 4,373 entities. Drawing insights from the literature and chosen scales, an electronic survey was crafted and disseminated via the Qualtrics platform from September 2021 to January 2022. Of the initial 216 acquired responses, a subsequent refinement yielded 202 valid entries. It is pivotal to acknowledge that the survey was tailored for the wineries' upper echelons, specifically targeting CEOs. Such an approach was rooted in the recognition that senior executives possess an enriched understanding of the enterprise's holistic and strategic operations, ensuring a profound grasp on the survey's dimensions. Consequently, the data reflects insights from 202 CEOs, each representing distinct wineries. These top-tier executives were tasked with addressing the construct components detailed in the subsequent subsection. Additionally, an alignment was discerned between the demographic characteristics of the broader cohort and the sample. The three autonomous communities predominantly represented in the broader dataset mirrored their prevalence in the sample. A comparable distribution was also noted between wineries engaged in wine tourism activities (57.43%) and their non-participating counterparts (42.57%).

Measurement of Variables

For the sake of ensuring the authenticity and dependability of the variables, previously validated scales were employed. Initially, wine tourism was gauged via a dichotomous metric, where a score of 1 denoted the winery's engagement in wine tourism activities and 0 indicated otherwise. Such a binary classification aimed to discern between wineries engaging in wine tourism and those that abstained, thereby discerning its potential influence on green innovation propagation. The emphasis was to ascertain the mere presence or absence of the activity, rather than quantifying its intensity. In defining the query aligned with this metric, Hall et al. (2000) provided the foundational blueprint, given its prevalent usage in wine tourism academic literature and its encapsulation of the activity's multifaceted attributes. Subsequently, the metric of sustainable performance was envisaged as a tri-dimensional construct, incorporating economic performance (spanning 4 items), social performance (comprising 5 items), and green performance (with 5 items). The adopted metric for sustainable performance was adapted from the combined works of Wang and Wang (2012), Paulraj (2012) and Paillé et al. (2014). Employed was a seven-tiered Likert scale, prompting participants to assess their institution's cumulative economic, social, and ecological performances vis-à-vis their market adversaries over the preceding five years. Furthermore, supplementary variables like organizational age, firm magnitude, and affiliation with Protected Designation of Origin (PDO) were integrated as modulating factors to discern their plausible effects on sustainable performance. PDO was conceived as a binary metric, where a score of 1 implied the enterprise's allegiance to at least one PDO, while 0 denoted a lack of such association. Conversely, the organizational age was gauged by calculating the span from its inception till the study's reference year (2022). It is imperative to note that the PDO and age queries took cues from the insights of Marco-Lajara et al. (2022b). As for the firm's size, the measurement framework of the Organization for Economic Co-operation and Development (OECD, 2005) was utilized.

Analysis Technique

The employed methodology to validate the posited theoretical framework was structural equation modeling (PLS-SEM) utilizing SmartPLS software (version 3.3.9). This method facilitates the exploration of intricate relationships amidst variables, some of which might be latent. Given the inherent nature of management research—where numerous constructs might remain intangible—such a technique proves invaluable (as acknowledged by Roldan & Cepeda, 2017). Specifically, the applicability of this technique is underscored by its capacity to gauge sustainable performance, a latent construct. Moreover, the research's sample size (n = 202) surpasses the stipulated minimum threshold of 100, deeming the technique's use as appropriate (as per Hair et al., 2019). Further bolstering the validity of PLS-SEM in this context is its efficacy in elucidating direct variable interrelationships, as well as its prior application in the wine industry's management studies, exemplified by works like Knight et al. (2019).

Results

Following the guidelines delineated by Hair et al. (2019), the findings are elucidated across three dimensions: (1) appraisal of the overarching model, (2) scrutiny of the measurement model, and (3) inspection of the structural model.

Initiating with the overarching model evaluation, it is discerned that the model boasts a Standardized Root Mean Square Residual (SRMSR) value of 0.048. This value is indicative of a commendable model fit, given that it is inferior to the 0.08 threshold postulated by Hu and Bentler (1998). Intriguingly, it even conforms to Byrne's (2008) more rigorous standard that necessitates an SRMSR beneath 0.05 for model endorsement. Subsequent to the validation against the SRMR benchmark, both the unweighted least squares discrepancy (d_ULS) and the geodetic discrepancy (d_G9) were ascertained to lie within the stipulated confidence intervals upon bootstrapping. Given that all resultant metrics

Table 1 Global model fit

	Value	HI95	HI99
SRMR	0.048	0.056	0.068
d_ULS	0.742	0.818	0.944
d_G	0.797	0.843	0.951

Source compiled by authors

are inferior to both HI95 and HI99, as delineated in Table 1, it can be inferred that the present model retains its credibility (Hair et al., 2019).

Pertaining to the assessment of the measurement model, several critical observations are registered. Firstly, the indicators associated with the variables under scrutiny affirm their individual reliability; this is evinced by their loadings, which surpass the 0.707 threshold as stipulated by Carmines and Zeller (1979). Furthermore, an internal consistency criterion is observed across all constructs, corroborated by both Cronbach's alpha and composite reliability values that eclipse 0.8. In addition, the constructs demonstrate convergent validity, as evidenced by the Average Variance Extracted (AVE) for the studied variables transcending the 0.5 benchmark (Fornell & Larcker, 1981). This implies that over half of the variance of the indicators is expounded by each respective construct. When subjecting discriminant validity to scrutiny, the Heterotrait-Monotrait (HTMT) criterion is employed, revealing that construct values invariably remain below the 0.80 mark. Such an outcome accentuates the distinctiveness of the constructs, validating their capacity to delineate discrete phenomena.

Thirdly, prior to delving into the analysis of the structural model, collinearity concerns were scrupulously inspected. Such a step was deemed imperative to circumvent potential multi-collinearity issues among the antecedent variables encompassed within the endogenous constructs. In line with the guidelines articulated by Hair et al. (2016), indications of collinearity manifest when the Variance Inflation Factor (VIF) exceeds 5. Consequently, VIF values transcending this threshold within the endogenous constructs indicate severe collinearity concerns. Yet, it was discerned that the VIF metrics obtained in this investigation consistently remained below this capstone. Referencing Fig. 2, insights into the R-Squared and β values, deduced from a bootstrap assessment with 5,000 subsamples, are delineated. Explicit examination of the direct impacts of wine tourism on economic (0.455), social (0.181), and environmental (0.112) facets revealed uniformly positive and statistically significant outcomes across all three dimensions. The computed R^2 values for economic, social, and environmental outcomes stood at 0.304, 0.262, and 0.149, respectively. Notably, each of these surpasses Falk and Miller's (1992) benchmark of 0.10, reinforcing the adequacy of the explained variance for each respective endogenous construct.

For assessing the model's quality, the Geisser test (Q^2) was employed, where values exceeding 0 ($Q^2 > 0$) were sought. Therefore, the model boasts a moderate predictive relevance, given that the Q^2 metrics surpass the 0.20 threshold as delineated by Kline (2011). In terms of control variables, the findings intimate a negative association between PDO membership and both economic and social performance. Contrastingly, the nexus between PDO membership and green performance emerged as positive and statistically noteworthy. In a similar vein, the company's age presented a negative correlation with the three performance dimensions, albeit the connection between age and social performance proved to be exclusively significant in its negative orientation. Meanwhile, company size showcased a positive and substantial association with all three performance modalities under scrutiny.

Discussion and Conclusions

The significance of this study is underscored as it empirically substantiates the positive and consequential association between wine tourism and the multifaceted performance—economic, social, and environmental—of Spanish wineries. Such tourism emerges as a novel avenue for the diversification of vintner establishments, thereby augmenting their competitive edge, fostering societal well-being within their operational domain, enhancing environmental stewardship, stimulating territorial progression, and bolstering their innovative capacities in an ever-evolving and challenging landscape.

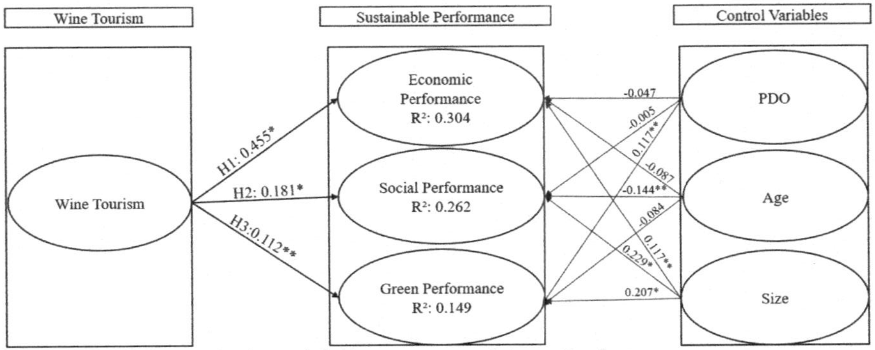

Fig. 2 Results of the proposed theoretical model (*Source* own elaboration)

This investigation provides insights into the nuanced impact of wine tourism activities on the triad of performance indicators. It is discerned that wine tourism exerts a profounder effect on economic performance, subsequently influencing social and environmental outcomes. Such differentiation can be attributed to the direct enhancements in business profitability wrought by wine tourism. Serving as a conduit for distribution, wine tourism amplifies direct wine sales at wineries, augments the profit margin per bottle by circumventing intermediary profit shares, promotes an array of products while spotlighting premium ones, and fosters direct client interactions, thereby cultivating brand advocates. This engenders a sentimental bond poised to translate into sustained wine sales. Such mechanisms undeniably bolster the economic resilience of wineries, ensuring their sustained viability. The findings herein resonate with the assertions made by Lavandoski et al. (2018), positing a favorable nexus between wine tourism evolution and organizational prowess in the Portuguese vinicultural landscape. Conversely, they diverge from the conclusions drawn by Smyczek et al. (2020), who did not identify a direct correlation between the depths of service provision in wine tourism and the intensity of direct sales in Italian vineyards. This intricate relationship between wine tourism and economic vitality in wineries necessitates further scrutiny to illuminate the academic realm of wine tourism. Given the existing scholarly contributions, it is evident that outcomes vary across viticultural nations.

In examining the nexus between wine tourism and social performance, a positive and substantial correlation between these entities is discerned. This alignment might be attributed to the enhanced conditions experienced by various stakeholders subsequent to the proficient execution of wine tourism initiatives. As previously articulated, wineries engaging in wine tourism pursuits potentially witness escalated profitability, which in turn might foster heightened job security and stability for their employees, given the fortification of their competitive stance. Such engagement not only augments job enrichment, by broadening the gamut of activities within wineries and offering avenues for specialization within the wine realm, particularly wine tourism, but also reverberates benefits to the larger community. Wine tourism emerges as an instrumental conduit for disseminating the rich tapestry of history, heritage, and gastronomic culture inherent to the wine-producing region, thereby amplifying its vinicultural identity. Thus, wine tourism adeptly addresses and caters to the aspirations and requisites of the myriad stakeholders who intersect with the operations of the winery. This research's insights harmonize with the findings of Poitras and Donald (2006), Andrade-Suárez and Caamaño-Franco (2020), and Frost et al. (2020), who underscore wine tourism's pivotal role in enhancing conditions for vineyard workers and its contribution to invigorating wine-centric regions through championing their cultural, historical, and heritage narratives.

The nexus between wine tourism and environmental performance, while positively and significantly correlated, manifests with subdued intensity. This nuanced relationship could be attributed to wine tourism's propensity to enhance environmental performance through more circuitous mechanisms. The infusion of ecological awareness among winery staff can be traced to several sources: (1) the collaborative dialogues within winery teams to adeptly convey their sustainable practices to the wine aficionados, (2) specialized eco-centric training curated for those orchestrating the wine tourism initiatives to bolster their environmental acumen, and (3) the proactive environmental inquiries or proposals tendered by the visiting wine enthusiasts. This augmented environmental cognizance subsequently translates into tangible enhancements in ecological performance, marked by judicious resource utilization (including electricity, water, and spatial resources) within wineries. Furthermore, these establishments might be inclined to adopt and showcase eco-friendly measures to magnetize a segment of wine tourists with pronounced environmental sensibilities. These observations resonate with the scholarly insights proffered by luminaries in wine tourism research, such as Montella (2017), Andrade-Suárez and Caamaño-Franco (2020), and Trigo and Silva (2022).

This investigation furnishes novel insights into the realm of wine tourism by collectively appraising the ramifications of such tourism on the economic, social, and environmental performance of wineries in Spain. Yet, while the contributions of this research are salient, it is imperative to acknowledge its inherent limitations. Specifically, the constraining nature of a cross-sectional research design is evident, as it precludes the examination of relationships across an extended temporal spectrum. Moreover, the gravitas of the subject at hand mandates a more expansive analysis encompassing other wine nations. To address these constraints, prospective research avenues may involve a longitudinal assessment of the surveyed firms and juxtaposing the articulated model within the Californian viticultural landscape. Such endeavors would facilitate discerning contrasts between the Old World and the New, rooted in the variables under scrutiny.

References

Alant, K., & Bruwer, J. (2004). Wine tourism behaviour in the context of a motivational framework for wine regions and cellar doors. *Journal of Wine Research, 15*(1), 27–37.

Anderson, K. (2013). Is Georgia the next "new" wine-exporting country? *Journal of Wine Economics, 8*(1), 1–28.

Andrade-Suárez, M., & Caamaño-Franco, I. (2020). The Relationship between industrial heritage, wine tourism, and sustainability: A case of local community perspective. *Sustainability, 12*(18), 7453.

Bruwer, J. (2003). South African wine routes: Some perspectives on the wine tourism industry's structural dimensions and wine tourism product. *Tourism Management, 24*(4), 423–435.

Byrne, B. (2008). *Structural equation modeling with EQS: Basic concepts, applications, and programming*. Psychology Press.

Carlsen, P. (2004). A review of global wine tourism research. *Journal of Wine Research, 15*(1), 5–13.

Canziani, B., & Byrd, E. (2017). Exploring the influence of regional brand equity in an emerging wine sector. *Journal of Wine Economics, 12*(4), 370–377.

Carmines, E., & Zeller, R. (1979). *Reliability and validity assessment (Vol. 17).* Sage Publications.

Carvalho, M., Kastenholz, E., & Carneiro, M. (2021). Co-creative tourism experiences–a conceptual framework and its application to food & wine tourism. *Tourism Recreation Research*, 1–25.

Charters, S., & Ali-Knight, J. (2002). Who is the wine tourist? *Tourism Management, 23*(3), 311–319.

Darnay, S. (2016). Quality of landscape and sustainability benefit to wine tourism: Contexts and commitments. In *Wine and Tourism* (pp. 15–25). Springer, Cham.

Duarte Alonso, A., Bressan, A., Kiat Kok, S., & O'Brien, S. (2021). Filling up the sustainability glass: Wineries' initiatives towards sustainable wine tourism. *Tourism Recreation Research*. https://doi.org/10.1080/02508281.2021.1885801,1-15

Falk, R., & Miller, N. (1992). *A primer for soft modeling*. University of Akron Press.

Festa, G., Shams, S., Metallo, G., & Cuomo, M. T. (2020). Opportunities and challenges in the contribution of wine routes to wine tourism in Italy–A stakeholders' perspective of development. *Tourism Management Perspectives, 33*, 100585.

Fornell, C., & Larcker, D. (1981). Evaluating structural equation models with unobservable variables and measurement error. *Journal of Marketing Research, 18*(1), 39–50.

Frigon, A., Doloreux, D., & Shearmur, R. (2020). Drivers of eco-innovation and conventional innovation in the Canadian wine industry. *Journal of Cleaner Production, 275*, 124115.

Frost, W., Frost, J., Strickland, P., & Maguire, J. (2020). Seeking a competitive advantage in wine tourism: Heritage and storytelling at the cellar-door. *International Journal of Hospitality Management, 87*, 102460.

Getz, D. (2000). *Explore wine tourism: Management, development and destinations*. Cognizant.

Getz, D., & Brown, G. (2006). Critical success factors for wine tourism regions: A demand analysis. *Tourism Management, 27*(1), 146–158.

Gómez, M., Pratt, M., & Molina, A. (2019). Wine tourism research: A systematic review of 20 vintages from 1995 to 2014. *Current Issues in Tourism, 22*(18), 2211–2249.

González, M. (2017). Wine tourism and sustainable environments. *Arbor-Ciencia Pensamiento y Cultura, 193*(785), 1–19.

Hair, J., Hult, G., Ringle, C., & Sarstedt, M. (2016). *A primer on partial least squares structural equation modeling (PLS-SEM)*. Sage Publications.

Hair, J. F., Sarstedt, M., & Ringle, C. (2019). Rethinking some of the rethinking of partial least squares. *European Journal of Marketing, 53*(4), 566–584.

Hall, C. (1996). *Wine tourism in New Zealand*, Proceedings of tourism down under II: A research conference (pp. 109–119). Donedin: Universidad de Otago.

Hall, C., Sharples, L., Cambourne, B., & Macionis, N. (2000). *Wine tourism around the world: Development, management and markets*. Editorial Elsevier.

Hall, C., & Macionis, N., et al. (1998). Wine tourism in Australia and New Zealand. In R. Butler (Ed.), *Tourism and recreation in rural areas* (pp. 267–298). Wiley.

Hu, L., & Bentler, P. (1998). Fit indices in covariance structure modeling: Sensitivity to underparameterized model misspecification. *Psychological Methods, 3*(4), 424–453.

Karagiannis, D., & Metaxas, T. (2020). Sustainable wine tourism development: Case studies from the Greek region of Peloponnese. *Sustainability, 12*(12), 5223.

Kline, R. (2011). *Principles and practice of structural equation modelling*. Guilford Press.

Knight, H., Megicks, P., Agarwal, S., & Leenders, M. (2019). Firm resources and the development of environmental sustainability among small and medium-sized enterprises: Evidence from the Australian wine industry. *Business Strategy and the Environment, 28*(1), 25–39.

Kotur, A. (2023). A bibliometric review of research in wine tourism experiences: insights and future research directions. *International Journal of Wine Business Research*, Ahead-of-print.

Kutateladze, R., Koblianidze, T., & Kochlamazashvili, L. (2021). Wine tourism as a mechanism for rural and regional development of Georgia. *Agricultural Economics and Rural Development, 8*(1), 17–29.

Lavandoski, J., Vargas-Sánchez, A., Pinto, P., & Silva, J. A. (2018). Causes and effects of wine tourism development in organizational context: The case of Alentejo. *Portugal. Tourism and Hospitality Research, 18*(1), 107–122.

Lorey, T., Dosquet, F., Durrieu, F., & Ambaye, M. (2023). Phygital innovations for wine tourism during the COVID-19 pandemic: The case of the jurançon wine cooperative in south west France. In *Technology advances and innovation in wine tourism: New managerial approaches and cases* (pp. 251–277). Springer Nature Singapore.

Marco-Lajara, B., Zaragoza-Sáez, P., Martínez-Falcó, J., & Ruiz-Fernández, L. (2022a). The effect of green intellectual capital on green performance in the Spanish wine industry: A structural equation modeling approach. *Complexity, 2022*, 6024077.

Marco-Lajara, B., Zaragoza-Sáez, P., Martínez-Falcó, J., & Sánchez-García, E. (2022b). Does green intellectual capital affect green innovation performance? Evidence from the Spanish wine industry. *British Food Journal*, Ahead-of-print.

Mitchell, R., & Hall, C. (2006). Wine tourism research: The state of play. *Tourism Review International, 9*(4), 307–332.

Montella, M. (2017). Wine tourism and sustainability: A review. *Sustainability, 9*(1), 113.

Nave, A., & do Paço, A., & Duarte, P. (2021). A systematic literature review on sustainability in the wine tourism industry: Insights and perspectives. *International Journal of Wine Business Research, 33*(4), 457–480.

OECD (2005). *Eco-innovation in industry: enabling green growth.* OECD.

OTRVE (2021). Informe de visitantes a bodegas y museos de las rutas del vino de España. Año 2021. [Report on visitors to wineries and museums on the Wine Routes of Spain. Year 2021]. Available at: https://wineroutesofspain.com/wp-content/uploads/2022/06/informe-visitas-a-bodegas-y-museos-rve-2021.pdf

Paillé, P., Chen, Y., Boiral, O., & Jin, J. (2014). The impact of human resource management on environmental performance: An employee-level study. *Journal of Business Ethics, 121*(3), 451–466.

Paulraj, A. (2012). Understanding the relationships between internal resources and capabilities, sustainable supply management and organizational sustainability. *Journal of Supply Chain Management, 47*(1), 19–37.

Piñeiro, M. V., de Salvo, P., & Giommi, F. (2019). Rural tourism and territorial development in Italy. In *Sustainability Assessment at the 21st century.* IntechOpen.

Poitras, L., & Donald, G. (2006). Sustainable wine tourism: The host community perspective. *Journal of Sustainable Tourism, 14*(5), 425–448.

Roldan, J., & Cepeda, G. (2017). *Modelos de Ecuaciones basados en la Varianza: Partial Least Squares (PLS) para Investigadores en Ciencias Sociales* (4a ed.). Universidad de Sevilla.

Serra-Cantallops, A., Ramón-Cardona, J., & Vachiano, M. (2021). Increasing sustainability through wine tourism in mass tourism destinations. The case of the Balearic Islands. *Sustainability, 13*(5), 2481.

Smyczek, S., Festa, G., Rossi, M., & Monge, F. (2020). Economic sustainability of wine tourism services and direct sales performance–emergent profiles from Italy. *British Food Journal, 122*(5), 1519–1529.

Sun, Y., & Drakeman, D. (2022). The double-edged sword of wine tourism: The economic and environmental impacts of wine tourism in Australia. *Journal of Sustainable Tourism, 30*(4), 932–949.

Toubes, D. R., Araújo Vila, N., & Fraiz Brea, J. A. (2021). Changes in consumption patterns and tourist promotion after the COVID-19 pandemic. *Journal of Theoretical and Applied Electronic Commerce Research, 16*(5), 1332–1352.

Trigo, A., & Silva, P. (2022). Sustainable Development Directions for Wine Tourism in Douro Wine Region. *Portugal. Sustainability, 14*(7), 3949.

Vázquez-Vicente, G., Martín-Barroso, V., & Blanco-Jiménez, F. (2021). Sustainable tourism, economic growth and employment—The case of the wine routes of Spain. *Sustainability, 13*(13), 7164.

Wang, Z., & Wang, N. (2012). Knowledge sharing, innovation and firm performance. *Expert Systems with Applications, 39*(10), 8899–8908.

Williams, K. (2021). Wine Tourism: From Winescape to Cellardoorscape. *Tourism Analysis, 26*(2), 245–248.

Zamarreño-Aramendia, G., Cruz-Ruiz, E., & Ruiz-Romero de la Cruz, E. (2021). Sustainable economy and development of the rural territory: Proposal of wine tourism itineraries in La Axarquía of Malaga (Spain). *Economies, 9*(1), 29.

Index

J. Martínez-Falcó et al. (eds.), *Wine Tourism and Sustainability*, https://doi.org/10.1007/978-3-031-48937-2

The manufacturer's authorised representative in the EU is Springer Nature Customer Service Centre GmbH, Europaplatz 3, 69115 Heidelberg, Germany. If you have any concerns regarding our products, please contact ProductSafety@springernature.com

Printed and bound by CPI Group (UK) Ltd, Croydon, CR0 4YY
01/10/2026
02225894-0002